Ethics in Engineering Design

Organizing Committee

E K Brodhurst
Institution of Engineering Designers

J R Lowe
Institution of Engineering Designers

C McMahon
Design Education Special Interest Group of the Design Society
and University of Bath

A Thomas
Design Education Special Interest Group of the Design Society
and Bournemouth University

C Dowlen
South Bank University

National Advisory Group

C Dowlen
South Bank University

C McMahon
Design Education Special Interest Group of the Design Society
and University of Bath

P R N Childs
University of Sussex

K L Edwards
University of Derby

M A C Evatt
Coventry University

C Ledsome
Imperial College

Ethics in Engineering Design

Proceedings of the
1st Design Education Special Interest Group of the Design Society
(DESIG) Annual Design conference and
10th International Conference on Product Design Education
10th–11th September 2003, Bournemouth University, Bournemouth, UK

Edited by

J R LOWE

Professional Engineering Publishing Limited
Bury St Edmunds and London, UK.

First Published 2003

ISBN 1 86058 413 6

A CIP catalogue record for this book is available from the British Library.

Printed by The Cromwell Press, Trowbridge, Wiltshire, UK.

About the Editor

Jean Lowe has worked as Education and Training Officer for the Institution of Engineering Designers following a career in further education and community education.

Related Titles of Interest

Title	Editor/Author	ISBN
Computer-based Design – Engineering Design Conference 2002	T M M Shahin	1 86058 372 5
Constraint-aided Conceptual Design (Engineering Research Series ERS 9)	B O'Sullivan	1 86058 335 0
Continuum of Design Education	N P Juster	1 86058 208 7
Design Applications in Industry and Education	S Cully, A Duffy, C McMahon, and K Wallace	1 86058 357 1
Design Management – Process and Information	S Cully, A Duffy, C McMahon, and K Wallace	1 86058 355 1
Engineering Design in the Multi-Discipline Era – A Systems Approach	P R Wiese and P John	1 86058 347 4
IMechE Engineers' Data Book – Second Edition	C Matthews	1 86058 248 6
Integrating Design Education Beyond 2000	P R N Childs and E K Brodhurst	1 86058 265 6
Sharing Experience in Engineering Design (SEED 2002)	M A C Evatt and E K Brodhurst	1 86058 397 0

For a full range of titles published by Professional Engineering Publishing (publishers to the Institution of Mechanical Engineers) contact:

Marketing Department
Professional Engineering Publishing
Northgate Avenue
Bury St Edmunds
Suffolk IP32 6BW
UK

Tel: +44 (0) 1284 763277: Fax: +44 (0) 1284 718692
E-mail: marketing@pepublishing.com
www.pepublishing.com

Contents

Related Topics

Authors' Index

Preface

Bournemouth University was pleased to host the International Engineering and Product Design conference 10th–11th September 2003. The IE&PDE 2003 Conference incorporated the 1st Annual Design Conference on Design Education under the auspices of the Design Education Special Interest Group of the Design Society (DESIG) and the 10th International Conference on Product Design Education under the auspices of the Institution of Engineering Designers. The venue was Bournemouth University which provided excellent accommodation.

The 'Ethics in Engineering Design' papers were categorized into sessions as follows:

- Curriculum
- Projects
- Related Topics

The papers presented in these proceedings were all scrutinized and refereed by the National Advisory Group to ensure maintenance of standards and relevance to the conference theme.

J Lowe
The Institution of Engineering Designers

Curriculum

QAA benchmark statements and Open University design courses

S ORGAN and **J DOYLE**
Faculty of Technology, The Open University, UK

ABSTRACT

A project was undertaken during 2002 to map a range of Open University courses onto the Quality Assurance Agency benchmark statement statements for engineering, to support the recently introduced Bachelor of Engineering with honours degree. The mapping exercise has encompassed courses from the faculties of Technology, Maths and Computing, and Science, and is seen as part of an on- going programme of quality control by the university. This paper is designed to explain the operation and outcomes of the project with particular reference to the two Open University design courses in those areas of the benchmarks covering professional and ethical responsibilities.

1. INTRODUCTION

1.1 The Open University

Before explaining in detail the nature, operation and results of the mapping exercise it is necessary to first outline the unique structure of the Open University (OU), its method of course delivery, qualifications and students.

The OU was established in 1969 and is now the UK's largest University, providing high quality supported open learning to over 200,000 students (1). The majority of OU students are mature students, studying on a part-time basis, often alongside full time employment. In common with other higher education institutions, the OU uses the Credit Accumulation and Transfer (CAT) framework to assign credit points to its courses: a year of full time University study equates to 120 CATs points. OU courses typically carry between 10 and 60 CATs points and can be combined in various specified combinations to make up a 360 point undergraduate degree. Each course is produced and managed by a course team, headed by a course team chair.

Named degrees such as the BEng (Hons) have closely defined specifications to ensure that the combinations of courses studied are appropriate both in level and in content. Nevertheless a large number of course combinations are theoretically possible; hence the importance of subject benchmarks to enable students to design a coherent, relevant and comprehensive degree profile. OU courses are specifically designed for distance learning and are therefore backed up by extensive published resources, including course information, teaching texts and assessment material. Such readily available and comprehensive information facilitates a mapping exercise of this kind.

The vast majority of students studying for an OU engineering degree will be actively working in the field, many of them with considerable practical experience of engineering and design.

1.2 The QAA Subject benchmark statements

The Quality Assurance Agency for Higher Education (QAA) was established in 1997 to provide an integrated quality assurance service for UK higher education (2). It is an independent body funded by subscriptions from universities and colleges of HE, its core business being: to review the quality and standards of UK higher education; to provide reference points, including subject benchmark statements; to advise government; and to manage validation of Access to HE. The Agency's mission is to promote public confidence that quality of provision and standards of awards in HE are being safeguarded and enhanced.

The QAA benchmark statements provide a means for the academic community to describe the nature and characteristics of programmes in a specific subject area. They represent general expectations about the standards of the award of qualifications at a given level and articulate the attributes and capabilities that those possessing such qualifications should be able to demonstrate. The benchmark statements for engineering courses are cross-referenced to SARTOR3, EPC Key Abilities and ABET 2000.

2. MAPPING OU COURSES

2.1 Subject and Skill Areas

The QAA benchmark statements for engineering list the skills, attributes and qualities expected of an engineer under six subject areas: mathematics, science, information technology and communication, design, business context, and engineering practice. The statements define four skills areas in each of the six subjects: knowledge and understanding, intellectual abilities, practical skills, and general transferable skills. There are three levels against which these skills can be measured: threshold, good and excellent.

For the purposes of the Open University mapping exercise a further element was added to specify whether the subjects were taught, developed and/or assessed (*tda*) in each course mapped. The initial mapping was carried out by John Doyle, working as an independent consultant, and was based on published information such as course descriptions, learning outcomes, course texts and assessment material. This generated sets of benchmark statements relevant to each individual course, which were sent to the course team chairs for comment, and modified where necessary.

In this paper we will concentrate discussion primarily on the benchmarks statements containing references to ethical and professional responsibilities. These appear in the "knowledge and understanding" section of the "business context" subject area. The statements at the three different levels are given in table 1.

The full range of benchmark statements can be found at http://www.qaa.ac.uk/crntwork/benchmark/engineering.pdf

Table 1. Engineering subject benchmark statements covering ethical and professional responsibilities

Knowledge and understanding of – Management of business practices (including finance, law, marketing, personnel and quality)	
Threshold	– *has a basic knowledge of management and business practices*
Good	– *has a good understanding of management and business practices*
Excellent	– *has an extensive knowledge and understanding of management and business practices, and their limitations*
Knowledge and understanding of – Professional and ethical responsibilities including the global and social context of engineering	
Threshold	– *is aware of the responsibilities of a professional engineer, and the ethical responsibilities of the engineer and also the impact of engineering practices in a global and social context*
Good	– *is aware of why an engineer must behave professionally and the responsibilities associated with working in and contributing to a multidisciplinary team* – *can analyse and present the ethical issues associated with a particular situation related to the discipline area*
Excellent	– *can explain specific examples of ethical and unethical, professional and unprofessional engineering conduct* – *can take a holistic view of engineering practice as part of a global society* – *can relate to ethical issues associated with situations outside the discipline area*

2.2 OU Design courses

Most OU Technology courses contain an element of design, and design is a major theme in the core engineering courses T173 Engineering the future and (new for 2004) T207 The engineer as problem solver. In addition, the faculty currently offers two courses specific to the subject:

T204 – *Design: principles and practice* is a 60 point course at second year undergraduate level which sets out to make the subject of design accessible and interesting to technologists and non-technologists alike. The main aims of the course are to develop:

- Design *awareness*, and foster an analytical judgement of designed objects by helping students to understand how decisions are made about the design of artefacts, the influences that contribute to those decisions, and the nature of the design process.
- Understanding of design *principles* applicable across a variety of professional practices such as product design, engineering and architecture.
- Design *skills*, by giving instruction, examples and experience in the use of basic techniques such as drawing, modelling and creative thinking.

T302 – *Innovation: design, environment and strategy* is a third level 60 point interdisciplinary course which asks why some innovations succeed while others fail to find a market. It tackles questions such as:

- Why has Britain a reputation for being good at invention but poor at exploiting new ideas?

- How does the British approach to innovation compare with that of other countries such as the United States, France and Japan?
- What is the most effective way of organising for technological innovation and managing innovative product development projects?
- What should be the role of governments in innovation?
- What skills and techniques do technologists, designers and managers need for innovation in the twenty first century?

The course explores such issues by means of case studies, theoretical readings and practical project work.

3. MAPPING RESULTS

3.1 Design courses

Tables 2 a) and b) show how the OU design courses T204 and T302 contribute towards a wide range of the QAA benchmark statements for engineering. Similar charts have been produced for other courses relevant to the BEng, and we hope to publish the complete set of data later in the year.

It is clear from table 2 that these courses can make a valuable contribution towards coverage of the engineering benchmarks in many areas. Both courses are particularly strong in design, and T302 in particular also scores highly in business context and engineering practice. Mathematics has been omitted from the charts since it is not specifically included in either course – however, engineering students taking these courses will inevitably make use of relevant mathematics during their studies.

To see how these courses equate to the benchmarks covering ethical and professional issues it is necessary to cross-reference the results in table 2 with the statements in table 1.

In mapping T204, the second level course, against the benchmark statements covering knowledge and understanding of business context it was found that the course met the criteria at the threshold level. This means that students successfully completing this course should: "*have a basic knowledge of management and business practices*" and " *be aware of the responsibilities of a professional engineer, and the ethical responsibilities of the engineer and also the impact of engineering practices in a global and social context*". By adding the additional Open University element to the QAA statements it was found that the course teaches, develops and assesses the above skills at the threshold level.

T302 – the third level course – was found to teach, develop and assess these skills at an excellent level. Students who successfully complete this course should therefore: "*have an extensive knowledge and understanding of management and business practices and their limitations*" and be able to " *explain specific examples of ethical and unethical, professional and unprofessional engineering conduct*" , " *take a holistic view of engineering practice as part of a global society*" and "*relate to ethical issues associated with situations outside the discipline area*".

Table 2. Results of mapping OU design courses against the QAA benchmark statements for engineering, showing the level at which each statement is addressed and whether the benchmark is taught (*t*) developed (*d*) or assessed (*a*).

a) T204 Design: principles and practice

	Knowledge and understanding	Intellectual abilities	Practical skills	General transferable skills
Science	THRESHOLD *t, d, a*	THRESHOLD *t, d, a*	THRESHOLD *t, d, a*	GOOD *t, d, a*
Information technology		GOOD *t, d, a*		
Design	EXCELLENT *t, d, a*	EXCELLENT *t, d, a*	EXCELLENT *t, d, a*	EXCELLENT *t, d, a*
Business context	THRESHOLD *t, d, a*			GOOD *t, d, a*
Engineering practice	GOOD *d*	THRESHOLD *t, d, a*	THRESHOLD *t, d, a*	GOOD *t, d, a*

b) T302 Innovation: design, environment and strategy

	Knowledge and understanding	Intellectual abilities	Practical skills	General transferable skills
Science	THRESHOLD d	THRESHOLD *d*		GOOD *d, a*
Information technology	GOOD *d, a*			GOOD *d*
Design	EXCELLENT *t, d, a*	EXCELLENT *t, d, a*	EXCELLENT *t, d, a*	EXCELLENT *t, d, a*
Business context	EXCELLENT *t, d, a*	EXCELLENT *t, d, a*		EXCELLENT *t, d, a*
Engineering practice	THRESHOLD *t, d, a*	GOOD *d*	GOOD *t, d, a*	EXCELLENT *t, d, a*

How were these findings arrived at? Firstly it must first be emphasised that these are **design** courses measured against benchmarks in **engineering** and as such against criteria they were not designed to meet, and the language of the course material does not therefore necessarily match the language of the benchmark statements. Secondly the statements are not confined to professional and ethical responsibilities alone but cover a range of management and business practices, so the "scores" allocated to each course should be interpreted in the light of this. However in examining the course material supplied to students on T204 it was found that, while there is not a specific unit dealing with ethical and professional issues, these themes run throughout the course. As the course title implies design *principles* as well as practice are taught, in some instances at a higher than threshold level as will be seen in the *design* subject area where the course is judged as teaching, developing and assessing these skills to excellent level. The basic course material is supported by descriptions and advice provided by professional designers, and students are made aware of how design is affected by many outside influences including economic and social changes and the nature of organisations for which designers work. The skills taught are developed through a series of design exercises, which include the critical examination of a consumer product from point of view of the user. As far as assessment is concerned there are seven Tutor Marked Assignments (TMAs) and an end of course examination. Generally speaking the issues are not explicitly examined, but it is clear from the nature of the assessment questions that students are expected to be aware of them.

Similarly in T302 professional and ethical responsibilities run throughout the course, and these issues are raised in the questions mentioned in the course description given above. However, as would be expected of a course at third level, in this course the issues are examined to a greater depth. The course shows why it has become increasingly important to consider the social and environmental consequences of design, and discusses moves towards the development of "greener" and "cleaner" processes. Skills are developed through a project which forms an important part of the course and which runs through almost the entire duration of study. Assessment of the skills in the subject is more explicit than in than in T204 and hence the course is found to meet the benchmarks as taught, developed and assessed at excellent level.

As outlined earlier Open University degrees and diplomas are made up of profiles, or combinations, of courses which in some cases are tightly controlled (e.g. the new OU BEng) but in others can be reasonably flexible. Both of the courses discussed above are compulsory components of the OU Diploma in Design and Innovation, so it can be seen that students gaining this qualification will have had a particularly good grounding in professional and ethical issues in design.

3.2 Other courses

In addition to the two design courses at least thirteen other current OU Technology courses contribute towards the business context / knowledge and understanding benchmark dealing with ethical and professional responsibilities.

The "systems" courses T205 *Systems thinking: principles and practice* and T306 *Managing complexity: a systems approach* are particularly good examples. The second level course T205 teaches, develops and assesses the issues at excellent level, the third level course T306 at good level. This seeming anomaly is a consequence of the benchmark language and of measuring courses against criteria they were not designed to meet. Some of the more mainstream engineering courses were also found to do well in bringing home to engineers and designers the importance of their professional and ethical responsibilities.

4. DISCUSSION

It must be stressed that the examples given above cover the results of mapping just a few courses against the benchmarks in one skills and one subject area. The entire exercise involved mapping some twenty two Technology courses in all four skills areas with each skills area having six subject areas; and as some mathematics and science courses were also mapped this gives an indication of the complexity of the project. The project was only made possible by the unique nature of the Open University teaching material available to us and is particularly relevant to the OU due to the flexible way in which courses can be combined to make up degree profiles. To date we know of no other similar exercise carried out in mainstream higher education, and would be most interested to hear of any such developments.

While it can be seen that mapping individual OU courses against the QAA benchmarks is not an exact science, in most cases a good match was found. It was also found in general terms that level one courses were consistent with the "threshold" level, level two with the "good" and level three with "excellent" in their specific subject areas. The exercise highlights the fact that the OU provides a range of courses covering all of the subject areas and skills included in the benchmarks at all levels, and therefore it is possible to develop many appropriate degree profiles from the courses available. A "typical" degree profile for the OU BEng could reasonably be expected to cover all the benchmark statements to at least threshold level, and to access higher levels in the majority of areas.

The bulk of the mapping is now completed, but the project will continue to be extended and updated as new courses become available. The results will be made available to BEng students as a planning tool, and are also likely to influence the design of new courses within the faculty and the possible development of themed course profiles. It is hoped that this work can be carried out in co-operation with the EC(UK) and the engineering institutions

5. CONCLUSION

It has been found that Open University courses from the Technology faculty contribute well towards the QAA benchmark statements for engineering, in terms of both content and level. In particular, the two specialist design courses teach, develop and assess a broad range of skills and include coverage of the professional and ethical responsibilities of an engineer, including the global and social context of engineering. Other courses, particularly in the systems discipline, also develop these issues. The mapping of a wide range of courses against the benchmark statements will aid students in selecting appropriate combinations of courses for the BEng, guide the faculty in providing comprehensive coverage of the engineering curriculum, and help professional bodies to review the achievements of students.

REFERENCES

(1) www.open.ac.uk
(2) www.qaa.ac.uk

Design for life-sustainable futures – are we all guilty?

R GRIFFITHS
School of Product and Engineering Design, University of Wales Institute, Cardiff, UK

ABSTRACT

This paper describes some the collaborative work undertaken between the School of Product and Engineering Design at the University of Wales, Institute, Cardiff, and Corus (formerly British Steel). The emphasis of our collaborative work has been on sustainable and ethical issues utilising the company's pre-coated steel and aluminium materials. This paper also explores some of the decision-making processes faced by product designers when specifying materials and processes for new products

1. A GUILTY CONSCIENCE

I am probably as guilty as the next designer by instinctively specifying plastics as the preferred choice of material when designing household products. This may also be true of many design undergraduates on our Product Design programme at the University of Wales, Institute, Cardiff.

Despite running modules that describe a variety of manufacturing techniques such as 'Design for Manufacture and Assembly' and 'Materials and Manufacturing Process Selection', students all too frequently identify a thermo-plastic polymer when creating and specifying new products. This raises the question of why this is the case when so many other materials and processes are available to them?

Research amongst the students concluded that it was a combination of factors that led them to select plastics as the preferred material of choice. Firstly, there is an image conjured up in their minds of what existing, competitor products looked like. This led to familiarity with the product type, which was invariably manufactured from an injection-moulded resin.

Secondly, their reaction is that the new product would be for mass-market consumption. Little or no thought was given to niche product marketing and consequently, injection-moulding techniques, ideal for mass-market consumables is deemed the most appropriate and cost-effective manufacturing process.

Thirdly, cost plays a major role in that students assume that cheap, high volume low cost production techniques are required for the majority of new products. They are often cynical about the loss of the Britain's manufacturing base in favour of third world countries like

China, Taiwan and Vietnam that employ child labour. They assume that in order for a product to be successful, it usually has to be cheap.

One can only conclude that long-term exposure to our consumer-led society has embedded an unhealthy approach to cost-effective sustainable design using finite natural resources. In the last two years staff within the School of Product Design have been making a deliberate effort to make students aware of current government packaging and environmental legislation and life cycle engineering issues to ensure a more responsible attitude is taken at the start of the product development process. After all, once Marketing has identified the need for a new product, it is the product designer that is responsible for creating a design that best embodies the performance characteristics. Supported by technical expertise, it is the product designer that usually determines the materials and manufacturing processes to be employed.

Design morality and sustainable design all too often are considered too late (if at all) by the current generation of new product designers. Why does the current generation of young designers only pay lip service to the needs of future generations? Is it the exception rather than the rule for new designers to consider end of life issues, or materials for their recyclability and ease of processing?

Product design cannot be undertaken in isolation. It is imperative to call on all available human and technical resources to develop innovative, fit-for-purpose value-added products. A Product Designer is truly a 'Jack of all trades'- creator, ergonomist, computer programmer, model maker researcher, materials processor, graphic artist, salesman and so on.

2. CORUS DESIGN COMPETITION

Recently, the School of Product and Engineering Design participated in a new design competition sponsored by Corus (formerly British Steel). The theme of this competition brief was to consider the sustainable issues when designing new products using their pre-finished 'Colorcoat' steel strip.

The Colorcoat range of strip steel materials offers a wide range of different colours, effects and finishes, many of which are not achievable with post-painting. In reality, you can develop unique surface finishes to suit any application. Steel can be laminated or painted to look and feel like stone, marble or even leather.

Corus's pre-finished Colorcoat steels are available in cut sheets and blanks, pierced and notched components and profiled sheets. The co-laminated coated steel is often used enclosures in high specification domestic appliances. The laminated films can be coloured and patterned to provide customers with a wide choice of uniquely designed finishes in small batch quantities. These are popular with interior designers for use in bars, nightclubs and other public spaces.

Corus offered the students extensive technical support from their New Product Development department and worked with them to develop bespoke, cost-effective precoated steel solutions for a wide range of applications. Corus have adopted three types of coating systems, which can offer a variety of colours, effects and performance characteristics:

- **Colorcoat** is a liquid paint applied system
- **Stelvetite** is a polymer film laminate bonded to the steel
- **Colorstelve** is a liquid applied system covered with a polymer film laminate.

The advantages to using these pre-finished materials include:

- They are cost effective to use as little or no post-finishing of the product is required.
- Materials that previously required painting and other chemical processes no longer require the purchase and storage of these items.
- The designer and specifier of these pre-coated steels can also demonstrate environmental responsibility.

Most domestic and industrial products can be designed using these materials regardless of operating environment or performance requirements. Pre-coated steel strip with anti-static and foodsafe surfaces can also be produced with ease.

The characteristics of these materials really opened the door of opportunity for our under graduates taking part in the competition. The brief was purposely constructed to offer as wide a scope of interpretation as possible. The brief, titled 'Environmentally Sound Furniture' read as follows:

'Create a piece of furniture for public or domestic use that delivers all the specified functionality, but which offers additional 'value', through being environmentally sound, while the piece also needs to consider social and economic requirements. The furniture piece must be principally composed of pre-coated steel and/or aluminium. It may derive its principal appeal from the use of an environmentally friendly material in a novel context, or for having being designed for the full life cycle, considering all the key environmental criteria from material selection through to end-of-life. It should exemplify efficient use of materials and resources and consider strategies such as reduction, reuse, refurbish and recycle.'

In reality, the word 'furniture' can be used to describe any piece of indoor or outdoor furniture. The design students were given sufficient scope to design anything from telephone kiosks and bus shelters to seating or shelving (Figure1).

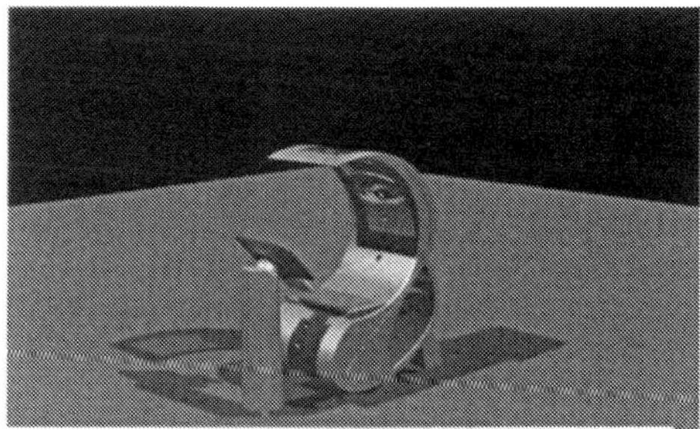

Figure 1. Interactive information booth : A project undertaken by Product Design Students at the University of Wales, using Corus precoated steels

Decorative information panels for utility boxes :A project undertaken by Product Design Students at the University of Wales using Corus precoated steels

Again I think many of us are influenced by our out of date perceptions of steel production. In a recent case study of an automotive suspension unit, new ultra-light steel components were actually 3% lighter than their aluminum alternatives and can be produced at a cost saving of 30%.

Peter Rawlinson of Corus Automotive Engineering commented: "The results explode the myth that the substitution of aluminium is the only answer to mass savings, and will help automotive engineers to better understand and exploit the latest steel technologies for more sophisticated and cost-effective mass-efficient designs."

On a domestic note, the use of steel is already widespread with applications in most product categories. For example, three out of every four cans we use at home are steel. We use thirteen billion steel cans every year and, rather worryingly, less than 8% of these, some one billion cans are recycled in the UK annually.

Like aluminium, steel is 100% recyclable and steel only uses about half the energy it takes to make an aluminium can. Corus is actively encouraging recycling in the UK by sponsoring various can collection schemes. These have proved successful to the point that 25% of every new steel can is made from recycled metal.

Students often fail to look beyond their initial design proposals to look at the investment requirements for their creations. I often stress the need to take a holistic approach to design – an approach in which we must consider all aspects of the product development process, from research, idea generation, prototyping, capital requirements and retail distribution and ultimately recycling and end-of-life considerations. By specifying thermoplastics and injection moulding, the naïve designer is, unknowingly committing huge capital resources to the manifestation of their design. Pre-coated steels on the other hand do not require this massive injection of capital resources and equipment to be produced and can be brought in and formed cost effectively.

The conflict between 'progress' and sustainability is a sensitive one. Mass consumerism and our appetite for ever-cheaper products often dictate the functionality, materials and processes employed to satisfy our hunger for gadgetry and a simplification of busy lives by utilising new technologies.

Metals in particular can play an ever-increasing role in the production of these products by using steel and aluminum – two fully recyclable products more effectively. Why use engineering thermoplastics, which have a finite life and are difficult and expensive to reclaim when 40% of 'new' steel is now produced from recycled resources? Laminated and painted plastics usually have to be pelletised and sent to landfill whereas painted and laminated steel and aluminium can be sent to the furnace for recycling.

Naturally, huge energy resources are still required to produce the raw steel and aluminium but I believe this is a necessary evil to satisfy our huge appetite for consumables.

3. WORLD RESOURCES

In a wider context, there are a number of current topical debates on environment issues and renewable resources of energy. Our response to these arguments highlights the apathy and selfishness demonstrated by society as a whole

For example, the government has announced that the UK is to get 18 new offshore wind farms as part of a plan to increase production of renewable energy. I have been following the arguments for and against the construction of a wind farm off Sker Point, near Porthcawl in South Wales. The proposal is that a 30-turbine development is placed four and a half miles offshore. No plans for any form of consultation with residents on its siting, size or desirability other than through the normal planning process are planned and herein lies part of our problem.

This wind farm will be a substantial development and will have a huge impact on Porthcawl and its tourism potential. It will also be visible from Swansea, Port Talbot and other towns along this coastline. The 30 turbines will be at least 300 feet high (higher than the Statue of Liberty) and will cover an area of at least 30 soccer pitches. But not all of this is negative. The company behind the £100m scheme, United Utilities claims the turbines will generate enough power for 86,000 homes during its 20 year life. Can it be counter-argued that the wind farm could also attract tourism? The local community is convinced that the proposal will have a detrimental affect on local tourism and have negative environmental effects including noise pollution and erosion of the local sandbanks affecting recreational activities.

Naturally the developer is quick to contradict these claims with its own counter arguments and benefits, but I believe the most important failure here is one of consultation and communication. Discussions with many local residents quickly expose a 'not in my back yard' syndrome and no consideration given to the long term benefits of wind farms or that the Kyoto agreement requires the UK government to produce atleast 10% of its energy from renewable sources by 2010. The opposition to these schemes fails to embrace the consequences of under committing to alternative and sustainable resources of energy.

This over-reliance on plastics to solve most of our problems fails to take into account the global economy and the threat of reduced oil supplies in future years. Pessimistic forecasters estimate that accessible oil supplies will run out within the next 38 years. It depends who you believe. In 1993, the world's available reserves were estimated to be just under a trillion barrels - about a 45-year supply of oil, based on current rates of consumption. The U.S. Geological Survey (USGS) argues that future supplies in West Africa, Russia, China and Canada ensure sufficient supplies of oil (some 2.1 trillion barrels) well into the next century.

Nevertheless, oil **is** running out – when will we start to embrace sustainable energy resources in time to minimise the impact of losing oil as one of the worlds leading natural resources? If we are to believe that there is no truth to the rumour that we are running out of natural resources, this is no excuse to become complacent about the prudent use of these resources. It is our responsibility to ensure that the next generation realise the consequences of an over-reliance on oil and all its derivatives.

We can all be responsible in playing a small role to change the mindset of the next generation to adopt and accept environmentally and socially acceptable energy resources as the default choice of sustainable energy, manufacturing process or material at the same time accepting that there will always be some negative arguments put forward by doubters. Ironically, these doubters will not be around when the problem manifests itself 50 years from now!

4. SPECIFYING MATERIALS

Research amongst undergraduate product designers into material selection revealed some disturbing responses. Despite being counseled on a plethora of production processes including most plastics manufacturing and metal forming techniques, the next generation of product designers remain steadfastly committed to designing products in plastic. In light of current government legislation on the environment this indicates that a lot more has to be done to alter the mindset of these students.

For example, when asked what materials would be suitable for the redesign of a smoke detector, all responded with an injection molded plastic. No thought had been given to suitable alternative materials nor market segmentation. For many new products to be successful, designers are increasingly attempting to identify niche market opportunities based on trends within small socio-economic groups and then pulling the rest of the market along.

When questioned about material and process selection recently, a student proposed a grade of ABS thermoplastic that was to be injection moulded. All very well if the product is for mass consumption and low manufacturing and retails costs were essential to market success. So how can alternative sustainable, recyclable materials be considered for this type of project? I shall return to market segmentation and the notion of exclusivity, albeit at a cost.

Material selection for today's young designers has never been easier. Whilst The British Plastics Federation and Dupont provide websites that can be interrogated for a plastic with the most appropriate performance characteristics, they do not exist to promote the environmentally responsible use of alternative materials. The Institute of Materials, Minerals and Mining provides a comprehensive advice and support service, and the Institution of Engineering Designers will answer all your technical queries - so why aren't we seeing greater use of steel, glass, and wood?

The ever-expanding range of manufacturing processes available to the designer can produce every conceivable form proposed by today's product designers. Rapid prototyping techniques such as rapid investment casting and laser cutting of steel sheet prior to forming can product complex three-dimensional forms. Appropriate material and process selection for the market is critical. Pre-coated steels are not appropriate for all applications but where selected in preference to thermoplastic resins, savings in tooling, processing equipment and waste are considerable.

Returning to our smoke detector, I hypothersised with the student –what would your smoke detector look like if style guru Sir Terence Conran or French designer Philippe Starck were to design it? I immediately got a different, more acceptable response – simplicity in design, glass, chrome, ceramics and wood became the materials of choice! With the exception of glass, the appearance of each of these materials can be replicated with pre-coated steels.

5. PACKAGING DIRECTIVES

Sustainable issues affect all aspects of the product development process. Frequently overlooked by the product designer is the design of the outer packaging whether it is for protection or promotional purposes.

It is easy to see why this element is not given greater emphasis when the UK government has already failed to implement a target reduction in waste materials. How can designers or the public be expected to adopt a culture of sustainability and responsibility for the environment when our own leaders fail to meet these targets?

How are we going to recycle or compost 30% of domestic waste by 2010? This amounts to around 27,000 tonnes of non-biodegradable, man-made plastic that must be disposed of in the UK each year. Landfill sites are becoming scarce and expensive. Legislation has already been introduced in Germany that states that the primary producer of a product is responsible for the disposal of its packing. This 'polluter pays' principle will be widely adopted throughout Europe and forward-looking manufacturers must now examine the recycling alternative.

At university I often discuss empowerment in the context of good business management and quality techniques. Cultural change has to adopted by all levels senior management if designers and engineers are expected to engage and embrace the current climate of sustainability.

6. CONCLUSION

Consumers and governments are making it clear that they want to deal with businesses that are environmentally and socially responsible. Today's product designers can play a key role in achieving these targets for environmental sustainability by working within all current legislation relating to protecting the environment. They can adopt a responsible attitude when designing products made from sustainable materials, cost effective and environmentally processes that meet the clients' performance criteria. They can demonstrate that it is possible to challenge consumerism and apathy.

So what of the smoke detector design? Through a process of brainstorming combined with thorough market research, we have identified a number of niche market opportunities for our client that can successfully integrate demanding performance characteristics with low cost, efficient and sustainable materials and processes.

The Corus environmental design competition concluded with the students proposing an exciting range of familiar domestic properties that exploited the performance characteristics of Corus pre-coated steels and aluminium. Product designs that incorporated unique aesthetics, features and benefits and fit-for-purpose solutions using environmentally friendly and sustainable materials. Indeed, we are currently protecting these designs with a view to commercially exploiting the intellectual property.

I can return to my role as a product design tutor confident that I am aware of current legislation, that I am aware of the benefits of selecting environmentally friendly materials and that I am prepared to pay for my clear conscience.

REFERENCES

(1) Timings RL, *'Engineering Materials'*, Pearson Education, 1999, ISBN 0-582-31928-5
(2) Swift KG, *'Process Selection'*, Arnold Press, 2001, ISBN 0-470-23774-0
(3) Corus Environment Report 2000

Websites

plastics.dupont.com – Dupont Engineering Polymers
bpf.co.uk – The British Plastics Federation

Incorporating life-cycle assessment within the teaching of sustainable design

C J BACKHOUSE, A J CLEGG, K G SNOWDON, and **T STAIKOS**
Wolfson School of Mechanical and Manufacturing Engineering, Loughborough University, UK

ABSTRACT

The paper describes the development of two teaching modules that support the concept of 'Engineering Design for Sustainable Development', with particular emphasis on the incorporation of LCA. The first module, entitled 'Engineering Design for Sustainable Development', was developed as an intensive, one-week module for a long-standing postgraduate MSc course in Engineering Design. The second, entitled 'Sustainable Product Design', was developed exclusively for the final (fourth) year of an MEng degree in Product Design and Manufacture. The paper summarises the teaching and learning experiences associated with the incorporation of LCA in the modules and provides guidelines for those teachers who would like to incorporate LCA in their courses.

1. INTRODUCTION

The Royal Academy of Engineering awarded a Visiting Professorship in Engineering Design for Sustainable Development to Loughborough University in 1999. Originally for three years, the RAE support has been extended for a further two years and so is currently in the fourth year of the five-year programme. The objective of the scheme is to encourage university engineering departments to appoint visiting professors who can transfer their industrial knowledge and expertise to the departments' teaching staff and students.

1.1 The Wolfson School

The Wolfson School of Mechanical and Manufacturing Engineering was formed in 2000 by the amalgamation of two established departments, Mechanical Engineering and Manufacturing Engineering, and makes up approximately one third of the Faculty of Engineering at Loughborough University. The School offers four undergraduate programmes:

- Mechanical Engineering MEng/BEng
- Manufacturing Engineering and Management MEng/BEng
- Product Design & Manufacture MEng/BEng
- Sports Technology BSc

The BEng and BSc programmes are of three years' duration and the MEng is of four years. Students can add an industrial placement year between the second and third academic years to gain an additional Diploma in Industrial Studies. The School has a current undergraduate population of approximately 805.

The School also provides taught postgraduate MSc courses in:

- Engineering Design
- Manufacturing Management
- Mechatronics
- Engineering Management
- Engineering Design and Manufacture

The School's postgraduate taught course population is approximately 40 full-time and 160 part-time students.

1.2 The Royal Academy of Engineering

The Royal Academy of Engineering (RAE) is committed to a long-term programme to encourage the effective application of engineering to improve the environment, to promote sustainable development and to protect natural resources. The RAE operates a highly successful scheme of Visiting Professors in Principles of Engineering Design that seeks to develop relationships between universities and engineers in outside organisations to enhance the learning experience for students. The success of this scheme led to the introduction, in 1998, of a scheme specific to Engineering Design for Sustainable Development. In 1999, Loughborough was able to appoint a Visiting Professor who held an appointment as Manager of Eco-design at Nortel Networks, a major multi-national, telecommunications equipment company. His objective was to work with the Wolfson School's academics to transfer knowledge and expertise to the staff and students.

1.3 Product Design and Manufacture

Product Design and Manufacture (PDM) is concerned with the creation of tangible and usable artifacts to meet a wide range of customer needs and requirements. It is the process by which ideas are converted into products. The programme is designed to provide industry with graduates possessing the skills and knowledge needed for the creation of successful world-class products. It is one of the few courses of its type to be accredited by both the UK Institution of Mechanical Engineers and the Institution of Electrical Engineers (Manufacturing Division). The subjects taught in Product Design are wide-ranging. As well as engineering science, design methodology and manufacturing, the topics of ergonomics and aesthetics in product design are covered. The theme of design is used to draw together subject areas and show their relevance in the creation of new products.

1.4 MSc in Engineering Design

The availability of well-designed products, processes and systems to meet the need of the market is the foundation of successful commercial enterprises. The programme in Engineering Design provides formal and practical education to meet the needs of the design activity in today's competitive markets. The objective of the programme is to provide a course of study that will enable the student to work effectively in an engineering design role, regardless of whether that role is concerned with the design of products, processes or systems at an overall or detail level. The programme consists of eight, week-long, taught lecture modules and project work. Each module is self-contained, and covers a complete design-related topic. The list of modules follows.

- Engineering Design Process and Project Management
- Engineering Design Methodologies
- Engineering Design Management and Business Studies
- Computer Aided Engineering
- Industrial Design and Human Factors
- Structural Analysis
- Materials Selection for Designers
- Engineering Design for Sustainable Development

Each student must also complete an individual project. Students are expected to commence the project in January and submit their report in August. During the period January to June the project work is undertaken concurrently with the module programme and the associated coursework and examinations.

2. SUSTAINABLE PRODUCT DESIGN

Sustainable Product Design (SPD) was developed exclusively for the final (fourth) year of an MEng degree in Product Design and Manufacture (PDM). This compulsory module was taught for the first time in semester 2 of the current academic year. Students registered for this module had already taken a module entitled 'Manufacturing for the Environment' in the third year of their course. The module was taught by a combination of lectures, case studies and a group project. Assessment was by a combination of a formal examination and the group project report, the latter constituting the coursework element for the module. Equal marks are allocated to the two components of assessment. The mark awarded to each student for the group project is determined by both an individual and a group component. The students were taught to use the Boustead LCA software tool and they were required to use it to support their selection of materials for the group project activity.

SPD deals with the elementary demand, essential product functions, the systems in which the products function; the nature, availability and selection of resources; and the distribution of those resources among nations and generations (1). It encompasses the concepts of ecodesign and design for the environment. However, SPD should be more than just environmental optimisation of products and services. It should also attempt to incorporate moral, ethical and social considerations because sustainability encompasses social and economic dimensions as well as resource conservation and the environment.

The module was developed to provide PDM students with an understanding of the tools and techniques available to facilitate SPD. It is also intended to provide knowledge of the product design processes that can reduce environmental impacts and promote sustainable practices. The contents of the syllabus include: SPD, Design for Environment (DfE), Life-cycle Assessment (LCA), quantitative and qualitative design guides, producer responsibility legislation, case studies, and project oriented activities. The group project ensures that the students apply the knowledge gained on a real industrial redesign project.

3. ENGINEERING DESIGN FOR SUSTAINABLE DEVELOPMENT

Engineering Design for Sustainable Development was developed as an intensive, one-week module for the Wolfson School's long-standing postgraduate MSc course in Engineering Design. However, the module is not exclusive to this course and accepts students from both within and outwith the School. The module operated in February 2003 with 42 registrations. The module is intended to provide students with an understanding of the environmental pressures acting on design to manufacture businesses and the scope that designers have to reduce the environmental impact of products throughout their life-cycle. The module consists of 35 hours of contact time that include industrial presentations, and case studies that are intended to provide students with an understanding of the practice of 'Design for the Environment'. Assessment is by examination and coursework with a mark distribution of 80% and 20% respectively. The module content includes: human impact on the environment; sustainable development; environmental legislation; materials and energy resource conservation; waste management; life-cycle assessment; design for environment; sustainable product design; case studies from the automotive, aerospace and white goods sectors; and business and environmental management issues.

4. LIFE-CYCLE ASSESSMENT

Life-cycle Assessment (LCA) attempts to assess the environmental impacts at each stage in the life-cycle of a product (2). LCA thus considers:

- Extraction and processing of raw materials.
- Manufacture of the product (and associated packaging and/or consumables).
- Use.
- End-of-life options (reuse/recycling/disposal).

By quantifying these impacts, LCA provides an objective basis for design choices between alternative materials/processes/products or systems. It is not suggested that the designer should conduct the assessment, but rather that the designer be provided with LCA output data that supports the choice between alternatives. An internationally agreed standard for LCA has been published by the International Organisation for Standardisation (ISO) and is documented in the ISO14040 series. The recommended methodology, shown in Figure 1, consists of four stages:

- Definition of the goal and scope.
- Life-cycle inventory analysis.
- Life-cycle impact assessment.
- Life-cycle interpretation.

4.1 LCA Teaching in Sustainable Product Design

LCA is taught in two formal sessions. The first is a two-hour lecture/tutorial combination that introduces and defines the methodology and requires the students to practice elements of the methodology. The second session is a two-hour demonstration/tutorial that introduces the students to the Boustead LCA software and then requires them to develop their skills in using the software for selected scenarios.

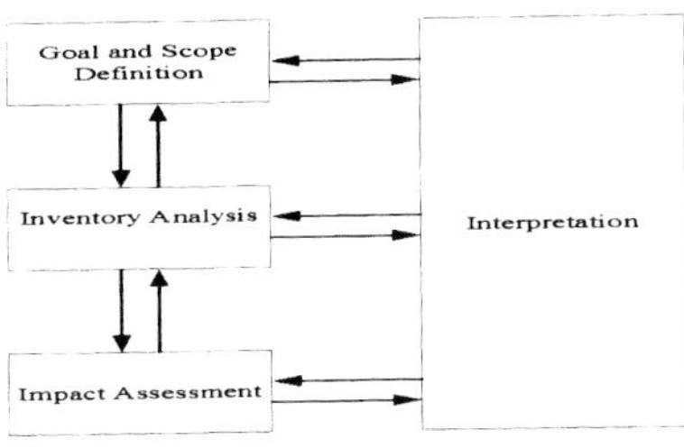

Figure 1: The LCA framework

The tutorial questions associated with the introductory lecture helped students to consider goal and scope formulation, functional units, flow diagrams and to compare the impacts of alternative choices (using data sets rather than software). Students worked in their project groups to conduct the tutorial exercises.

The steps in the Boustead LCA software are:

- Construct a reliable and detailed flow diagram of the sequence of unit operations that form the system of interest.
- Collect input and output data for every unit operation.
- Enter the data into the Boustead model.
- Calculate the data.
- Read the data.
- Determine values for environmental effects.

Following the introduction to the Boustead software the students gained experience in its use by considering a number of scenarios. These included oxygen production in the UK; bauxite mining in Australia; production of 1 kg of dry sand in the UK; silicon carbide production and an industrial process in the UK.

These skills must subsequently be applied in DfE group projects in which the students must demonstrate the use of the LCA software to substantiate a choice between alternative materials. The students work in groups of three to undertake the DfE project. In the current year the artifacts considered were supplied by Jaguar Cars Ltd. and a representative from the Company briefed the students on their task. The artifacts consisted of a laminated fuel tank with appendages, an external bumper cover/assembly and a sun visor.

The students were asked to provide feedback on the LCA software training activity by scoring from 1 to 5 the four aspects shown in Table 1. The table also shows the overall average response score.

The Boustead LCA software was available within the School because it had been purchased to support a research contract. The cost of the software, the licence to operate it on four machines and the training totalled approximately £2250.

Table 1: LCA activity feedback

Aspect	Score
New knowledge gained	4.44
Intellectual challenge	3.78
Interest value	4.56
Enjoyment value	4.00
Overall average response	4.20

4.2 LCA Teaching in Engineering Design for Sustainable Development

Timetabling constraints required that LCA be taught by a combination of a formal introductory lecture complemented by a case study. This section briefly describes the LCA case study that was based on Nortel Networks' experience in the assessment of LCA software packages. It requires that students evaluate the alternative choices of aluminium alloy or polymer for use in the faceplate (front cover) for a rack containing telecommunications equipment.

4.2.1 Nortel Networks Faceplate LCA Case Study

The case study evolved from a Nortel Networks investigation to evaluate two software tools for conducting LCA. The software tools were the Boustead LCA software and the PIRA International Model. 'Look and Feel' panels were selected for the LCA exercise. These panels are used as faceplates for the sub-racks of equipment for printed circuit board support, shielding from electromagnetic and radio frequency interference, and structural integrity in Transport Node Switches. Two material choices were considered: aluminium alloy and plastic. The faceplates were chosen because they represent a relatively simple part in terms of construction and materials.

The functional unit chosen for the exercise was a 16 inch (406.4 mm) wide sub-rack of equipment with an appropriate number of panels filling the space. The number and size of panels for a standard sub-rack was established and this mix of panels was specified as the functional unit. This produced total weights of 967.9 g and 686.7 g for aluminium and plastic respectively. The aluminium panels were produced as extrusions and the plastic panels were injection moulded from a polymer blend with a phosphorus compound flame retardant and titania and carbon black pigments to produce a final grey colour. The plastic panels were selectively electroless plated on their inner surfaces. An example of a plastic panel is shown in Figure 2.

4.2.2 Nortel Networks Experiences in the Use and Limitations of LCA

Databases are still evolving to incorporate new data as it is generated for specific processes. If an exact fit is not possible, a compromise may be necessary. For the LCA tools used, it was necessary to treat the aluminium as ingot rather than an extrusion. Similarly, as the data for the particular polymer blend was not available in either database, an alternative was specified for the purpose of the exercise.

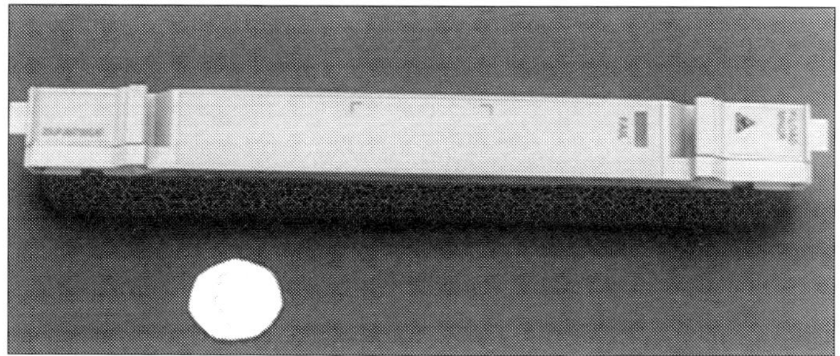

Figure 2: An example of a plastic faceplate component

5. EXPERIENCES

The undergraduate students had received an introduction to LCA in the compulsory third year Manufacturing for the Environment module. However, the SPD module required that they develop expertise in both LCA methodology and in the use of LCA software. The SPD module activities were intensive but appropriate to able MEng students. It is not suggested that this relatively short introduction enabled the students to master the complexity and detail of the Boustead Model and software. However, the requirement to use the software during their group project phase extended the students' experience in the understanding and use of the LCA software. This phase was complemented by extended tutorial support from a member of staff skilled in the use of the software. The size of the group is clearly important in terms of the commitment of resources and availability of support. With a module group of only nine students, this was relatively easy to manage.

The undergraduates' responses to the simple feedback form provided a positive outcome, although it is interesting to note that the 'intellectual challenge' component received the lowest score. This may provide the opportunity to introduce tutorial questions and scenarios that are more complex than those currently provided. However, it will be interesting to see if this disdain for the intellectual challenge extends to the LCA component of the group project. The students may find the challenge of dealing with the complexity of a real industrial problem rather more demanding than constructed tutorial questions and scenarios.

The time available to present the LCA topic to the MSc group was somewhat less than that available to the MEng students and the former did not have the benefit of an introduction to the topic in an earlier module. The approach adopted for the MSc students included an introductory lecture followed by an industrial case study to demonstrate a practical experience in the use of LCA. This approach enabled the benefits and limitations of the LCA approach to be successfully demonstrated.

In offering a guide to lecturers who might like to include LCA in their teaching of engineering/design topics, the approach proposed would depend on whether or not LCA software was available. Even without such software it is certainly possible to describe the principles and practice of LCA using the combination of a formal lecture, tutorial exercises and a case study. This was the approach adopted for the large MSc group for whom an overview of LCA was important within the context of the module objectives. For those students of the MEng course, for whom a detailed knowledge of DfE tools was required, the opportunity to develop skills in the use of LCA software and then apply those skills to a real industrial problem provided an additional dimension to their learning opportunity.

6. CONCLUDING SUMMARY

The paper has described how life-cycle assessment is being incorporated within two modules that support the teaching of environmental and sustainable development principles within the Wolfson School's undergraduate and postgraduate course portfolios. LCA is an important tool for quantifying certain environmental impacts and for providing the basis for an objective comparison between alternative choices. However, LCA tools are expensive to purchase and operate and they have limitations. It was not intended that the teaching of LCA described in the paper produce skilled LCA practitioners but rather that the students become aware of how LCA is conducted and the limitations of the process. It is most likely, certainly in larger organisations, that designers would be provided with DfE tools that suggest preferred choices of materials. Those choices would have been determined by LCA conducted by specialists either within the organisation or based in external consulting organisations.

7. ACKNOWLEDGEMENTS

The authors wish to acknowledge the financial support of the Royal Academy of Engineering and the financial and practical support of the Directors of Nortel Networks.

REFERENCES

(1) Lewis, H. and Gertsakis, J., *"Design + Environment: A guide to greener goods"*, Greenleaf Publishing, Sheffield, 2001.

(2) Charter, M. and Tischner, U. (eds), *"Sustainable Solutions: Developing products and services for the future"*, Greenleaf Publishing, Sheffield, 2002.

The education of an informed and critical ethical designer

A DEAN and **R PARKER**
Division of Design, Technology, and the Built Environment, University of Derby, UK

ABSTRACT

Ethical design is based on the desire to minimise the environmental and social consequences of products in the broadest sense. It is recognised that all engineers and designers should be aware of the issues concerned and be able to design eco-friendly products. This paper describes the use of a number of teaching methodologies that have been developed over the last few years to enable students to gain an appreciation of wider social issues, the science behind the ecodesign hype and the latest computer tools used by business to ensure they produce eco-products. The methodology also challenged current ideology that a westernised industrial society is the only model on which to base the planet's future.

1. ETHICAL DESIGN IN CONTEXT

1.1 Introduction

'Ethical' may be considered to be synonymous with sustainable for many designers: both terms are interchangeable and concerned with social, environmental and economic issues (1). The value of the formation of the ethical designer as a distinct entity is being increasingly recognised in the higher education sector. The global consequences of the design, manufacture, use and disposal of all products has moved into the mainstream and both businesses and individuals have to accept their social responsibilities. This means that business is more ready to engage with this area and demand ethical designers as prospective employees (2).

At Derby we offer a range of programmes of undergraduate study in design, all of which engage in ethical debates. However, both the BSc(Hons)Product Design (Innovation and Ecodesign) (PDIE) and the BSc(Hons)Architectural Technology and Digital Innovation (ATDI) programmes offer a more in depth insight into these areas, in their respective disciplines. Both sets of students use taught theory to design products that manipulate raw material inputs, manually or mechanically, into finished articles. The programmes also educate students to assess other inputs and outputs such as indirect inputs of physical resources (buildings, factories) and mental resources (awareness, management) and outputs, some beneficial to society (the product itself) and some not (pollution).

This paper seeks to explain how lack of awareness in students of the negative impacts produced during the lifecycle of a product can be addressed as part of the normal curriculum. Ethical design is emotive and previous to the present initiative was not viewed with any great sense of personal ownership or understanding on the part of students and academic staff, but rather as an academic exercise which did not affect their daily lifestyles or actions. The authors aim, through this paper, to describe how this perceived deficiency was addressed in all stages of the degree programmes above, but how primarily at stage 1 underpinning knowledge and critical evaluation were introduced to support the succeeding stages.

Ethical design is based on the desire to minimise the environmental and social consequences of products in the broadest sense. In order to achieve this, graduates must be armed with a variety of knowledge and skills that cross the boundaries of traditional subject disciplines such as design, science, humanities and engineering. They must also possess critical evaluation skills that allow them to assess a number of different eco-friendly alternatives and to make an informed decision.

1.2 The global situation

The argument for embedding knowledge about the environment in students and the need for resource awareness are reinforced by consideration of global and domestic weather patterns. The 1990s were the warmest decade on record, 1998 the hottest year on record and the worst for storm damage in the UK (3). Furthermore, Autumn 2000 was officially the wettest since records began in 1766, with an average 457mm of rain falling between September and November (4). Globally, 2002 was predicted to be the second warmest year since 1860 (5). A pattern seems to be emerging and these statistics form the basis for the interest underpinning the climate change agenda. Indeed, there is a consensus view of the world's leading climatologists, supported by business leaders and politicians that a positive link between global warming and increased fossil fuel usage does exist (4).

As Figure 1 identifies (6), the rate of consumption of fossil fuels since the middle of the Industrial Revolution has significantly increased particularly from post World War 2. This is a period characterised as the era of 'mass consumption'.

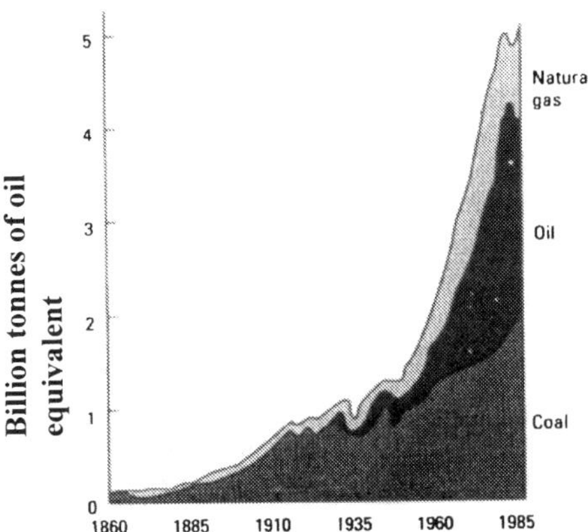

Figure 1 – Fossil Fuel Consumption 1860 - 1985

Students are introduced to the dichotomy that says an Indian's annual consumption of commercial energy is the equivalent of 210 kg of oil, whilst an average UK consumer uses the equivalent of 3,756 kg over the same period (7). With the predicted population in India set to increase from 1008.90 millions to 1572.1 millions in 2050 (8), together with a demand for human rights for Indians to have a similar lifestyle as a person from the West the students are asked to consider the problem from both the developed world's viewpoint and the Indian's.

Waugh has suggested the approximate dates by which a country reaches a stage of economic development (9). Table 1 appears to confirm Elliot's findings above, that the period (Stage 5) from 1930s onwards, saw significant consumption of fossil fuels.

Table 1 – Relative stages of economic development
Key to table:

- Stage 1 = Traditional society (assumed all countries meet this criteria),
- Stage 2 = Infrastructure in place,
- Stage 3 = Commencement of economic exchange with foreign countries, development of institutions to support society etc.,
- Stage 4 = Drive to maturity,
- Stage 5 = High mass consumption.

Country/Stage	**2**	**3**	**4**	**5**
UK	1750	1820	1850	1940
USA	1800	1850	1920	1930
Japan	1880	1900	1930	1950
Venezuela	1920	1950	1970	N/A
India	1950	1980	N/A	N/A
Ethiopia	N/A	N/A	N/A	N/A

Predicted population increases in the Far East and other third world areas which are entering the period of mass consumption will add considerable strain to fossil fuel demand and climate change. Students are asked to examine this paradox between the western, developed lifestyle compared with the developing world.

The authors consider that the developed world has a duty to redress consumption balance and take the lead in ethical design. However, this will be a long-term objective and delivered to successive generations through the medium of higher education.

1.3 The personal situation

The global situation presented, does lead to the conclusion of a world out of control. Despite attempts at harnessing sustainable and ethical consumption for all, initially through the Brundtland Commission, then the Rio and Kyoto summits, it appears that these global institutional frameworks are not working. Through Agenda 21, it was considered that sustainable patterns of society could filter down with the local government playing a key role in its vision.

In Europe, a survey into public opinion on environmental issues in the European Union discovered a high level of apathy and unawareness of key issues. However, a 44% response was elicited when people were questioned about whether they believed that changes in their everyday lives, in areas such as transportation and recycling, could reduce environmental damage (11). This concern is not put into practice in the UK - we have one of the lowest rates of recycling waste within Europe (12).

Although students are generally interested in using eco-friendly materials, many lack a sense of awareness of how the rise of western consumerism has affected the stock of materials and the environment (13). In the architectural/construction field, a recent report found that a number of developed and developing countries were concerned with the relatively slow pace with which innovations contributing to sustainable development and construction were being adopted (14). Thus, the gulf between the graduate and businesses' social responsibility can be vast in terms of knowledge base.

The authors considered an appropriate teaching methodology to bridge this gulf would be to examine the ethics of the students' daily lives, primarily their time and consumption patterns outside the learning environment. Students were asked to make a diary of their day, the environmental consequences of their activities and to suggest more eco-friendly alternatives. During the first semester, in a personal skills module, students from PDIE produced a website entitled "Student Guide to Eco-Living". Each student was responsible for a 'page' within the site that reflected their interest in the subject and a menu page that provided an overall theme to the work.

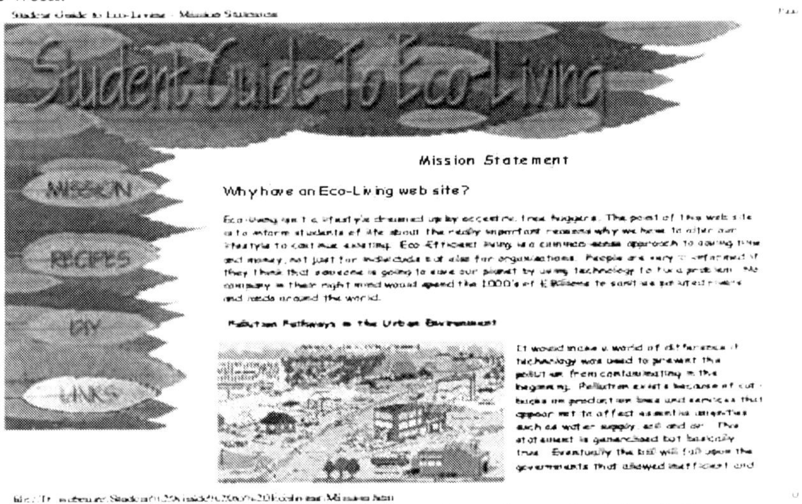

Figure 2 – Example of menu page from the students' website

1.4 Bridging the global and personal at stage 1

Many graduates from the PDIE and ATDI programmes will be working on the design and production of products ranging from pens to large multistorey buildings. However, the aim of the authors is to imbue critical ecologically friendly principles to underpin an individual's professional work, irrespective of the work situation and the products they will produce. We also hope that the graduate will engage in the same way with society and not restrict their involvement to the products they design. The aim is to arm students at stage 1 with a critical awareness of the science and legislation for the four aspects of ecology: air, water, noise and waste that affect ecodesign. In the module Industrial Ecology, students are required to complete a 'Green Design Project' The assignment assesses the learning objectives: be knowledgeable on current environmental legislation affecting industry and what affect this has on companies, be aware of the ethics involved in designing for the environment with the view to developing an environmentally sounder approached to design

The students are given the brief that they are part of an environmental design group working for a large multinational car manufacturer that is responsible for advising the company on environmental issues that affect their products. The design team is in separate parts of the world and has been asked to work together to produce a design concept for a new energy efficient family car. The team is required to produce environmental information and a possible green concept design car that will need to be taken into account when the designers produce the final design. This information will not only be to the legal requirements the company have to meet, but also any green design strategies that could be incorporated into the design that would give a marketing 'edge'. It would also include the moral issues faced by the company and how the public may react to the product. The website front page produced by one group of students is shown as Figure 3 and the Eco Issues page as Figure 4.

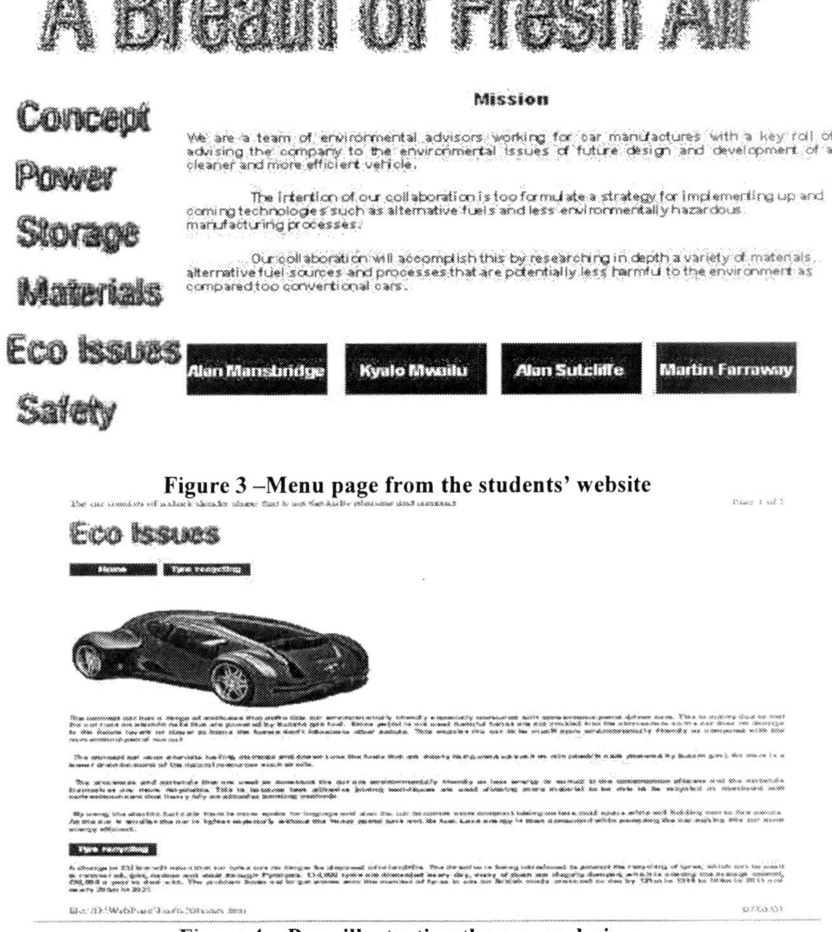

Figure 3 –Menu page from the students' website

Figure 4 – Page illustrating the ecocar design

However, in some programmes the concern for the environment seems to be outweighed by other considerations. A typical example illustrates our concern:

A presentation given by stage 3 BSc(Hons)Product Design, Innovation and Marketing students. They delivered a presentation to their peer group about hydroelectricity as a power source. When questioned as to whether the installation of a dam would affect the local communities downstream, the students' response was that the locals had no right to stand in the way of progress and should move somewhere else.

Where appropriate, examples such as the one illustrated above are passed to students for information during lecture periods or informal meetings in order to underpin the importance of ethical behaviour. Despite agreeing in principle with ethical considerations, their personal behaviour, certainly outside the classroom in terms of time and consumption patterns remains firmly unecofriendly in nature.

Students behaviour will certainly change in the long-term, but there is much that universities can and should do to embed a critical awareness of personal behaviour beyond academic study to embrace all aspects of lifestyle. Current provision in secondary schools is for students to be aware of the overall issue surrounding climate change etc but the link needs to be reinforced at Higher Education level.

1.5 Developing skills in ecodesign in stages 2 and 3

The module Environmental Use of Materials can illustrate the development of the students' knowledge and understanding within stage 2. The module is dedicated to one product, a caravan (ay 2002/3), through which the students learn how to modify the design, by the use of different materials, in order to produce a 100% recyclable product. Basic materials concepts are developed within this module to include the selection of materials using the Cambridge Engineering Selector (CES) and the use of a simple lifecycle analysis technique. An example of some of the graphs used to select a suitable environmentally friendly material for the internal fixtures in the caravan is shown in Figure 5.

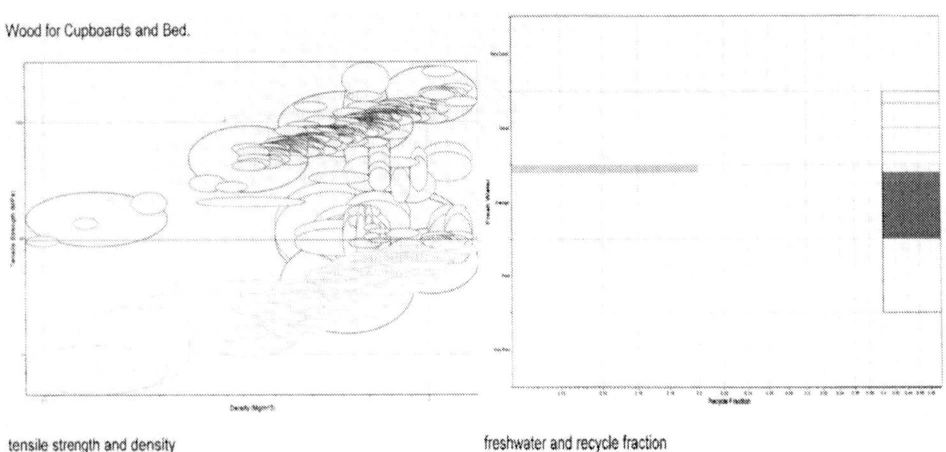

Figure 5 – Cambridge Engineering Selector graphs

The stage 2 module Ecodesign Strategies and Methodologies takes a much more analytical view of the subject and allows students to use Eco-It and Simapro to qualitatively and quantitatively assess the environmental impact of existing products. Products may then be redesigned to reduce their environmental impact. Students produced lifecycle analyses for a conventional upright vacuum cleaner and a cyclone cleaner and compared the impacts. They then offered a redesign to reduce the impacts. Students were asked to act as though they were part of a development team reporting to the managing director of a large company manufacturing vacuum cleaners. The company has two types of cleaners it manufactures - traditional and cyclone. They were part of strategy review that is assessing all the company's products and recommending how the company can progress its product portfolio over the next 10 years. They were supplied with two cleaners one traditional and a new Dyson cleaner. Students produced reports and presentations based around the phases shown in Table 2. Students assessed each other's work.

Table 2 – Phases in analysis and development of an eco-product

Phase 1	Product Selection and General Product Analysis Market Competing Products Pressures and potential for a change of product Product information Process Tree with possible supplied products
Phase 2	An analysis of the products' environmental impact and setting of the design directions. Life-cycle analysis (using Simapro) Process tree Limited inventory analysis Impact analysis matrix
Phase 3	The realisation of a new environmentally improved product. Redesign Marketing Incorporating redesign into your company

The use of the Independent Studies module as a means of connecting all the ideas within the degree has been used successfully for a number of students. Projects including the design of shoes for the third world, an ecofriendly board game for children and the use of heat from waste to heat greenhouses are just some of the successful projects by different students.

A strategy to heighten ethical awareness is being extended across the programmes. Thus the initiative is now embedded within the stage 1 module, Materials Technology for Design, which is offered on both the architectural and product design programmes. Succeeding stages offer modules such as Appropriate Technology where students are asked to join a current local environmental debate. In the last academic year this was the siting of a recycling and incineration plant in Sinfin, Derby. Students role played interest groups such as developers, local residents, the local council and environmental groups. The 'public enquiry' held by the students was attended by the developers, local interest groups such as Friends of the Earth and the Derbyshire Renewable Energy Centre.

2. EXTENSION OF THE PROJECT

From the staff and students' experience of the pilot study, the initiative was successful in that it demonstrated that staff were committed to educating students regarding ethical issues.

2.1 Teaching and Learning Strategy

Staff feel that the teaching and learning strategy used engaged students and produced commitment in the modules and assignments. For the next academic year it is envisaged to extend the project as below. The use of IT equipment within the project enhances the students' experience and introduces a 'fun' element into the modules.

Internal debate amongst teaching staff concerning the module elicited the following comments.

Time - We need the student to be able to record how they use their time throughout a day. What are they doing?

Consumption - Are students aware of the consequences of purchases they make? Again, a record needs to be kept.

Record production - The use of the Internet to record students' time and consumption patterns in the form of a 'Video Diary' is suggested for next year. This builds on the pilot study and we believe is a highly engaging form of record keeping.

Resources - The University has extensive digital equipment which can loaned to students as needed.

Underpinning knowledge - Students on entering the programme have differing levels of understanding of sustainable/ethical issues and this needs to be addressed early in the course.

REFERENCES

(1) Hockerton Housing Project *'The Sustainable Community: A Practical Guide'* pp 6.
(2) Elkington J. *'Cannibals with Forks'* (1999) Capstone pp 5.
(3) New Statesman 6/3/2000.
(4) Environment Action December 2000.
(5) http://news.bbc.co.uk/1/hi/sci/tech/2583357.stm
(6) Elliot D. *'Energy,Society and Environment'* Routledge (1997) pp 4.
(7) Tully M. *'No full stops in India'* (1992) Penguin pp 9.
(8) New Statesman 4/11/2002.
(9) Waugh D. *'Geography - An integrated approach'* (2nd Edition) Nelson (1995) pp 574.
(10) Rifkin J. *'The Hydrogen Economy'* Polity Press (2002) pp 24.
(11) http:www.greenconsumerguide.com
(12) http://news.bbc.co.uk/1/hi/uk/1668242.stm
(13) Maynard G. *'We're Rubbish at Recycling'*, Daily Express, Monday 6th January 2003.
(14) International Council for Research and Innovation in Building and Construction (CIB) quoted in Graham P. *'Building Ecology'*, Blackwell.

The challenges of introducing sustainable development principles into product design teaching

G HOWARTH
Smith and Nephew Group Research Centre, Heslington, UK (Retired)
M HADFIELD
School of Design, Engineering, and Computing, Bournemouth University, UK

ABSTRACT

It is vital that Engineering Product Designers understand and apply the principles of sustainable development in their university courses and careers. The Royal Academy of Engineering has funded a project to introduce case study teaching in sustainable development design issues to undergraduate students at Bournemouth University. The scheme allows technology transfer from business and industry. The industrial link is provided by Smith & Nephew plc; a major International Medical Device manufacturer. This paper addresses the specific educational challenges of student-centred web-based learning and teaching strategies applied to the complexity of sustainable development and the needs within industry. The approach has been received enthusiastically and will form the basis for further developments.

1. INTRODUCTION

1.1 Industrial Perspectives

There are many pressures on UK business today to respond to the UK Government's strategy for sustainable development, a Better Quality of Life (1). Similar priorities are now being identified at European level, with the emergence during 2001 of the Community's Sustainable Development Strategy, White Paper on Corporate Social Responsibility and the Sixth Environmental Action Programme (2,3,4). The headline priorities in those strategic Community documents are:

- Limiting climate change
- Managing natural resources more responsibly
- Addressing threats to health and the quality of life

More recently the Johannesburg Summit recommended much more focus on the effecient use of resources and in particular the consumption and production of products. This has resulted in the UK Government preparing a UK strategy for sustainable consumption and production (5)

One way of demonstrating a response is for a company to report on its Sustainable Development activities, usually following the Global Reporting Initiative (GRI) guidelines (6). Smith & Nephew plc has produced earlier this year a third Sustainable Report 2003 (7), which does commit to carrying out more measurements in all areas, but in particular to develop an approach to Sustainable Product Development. The philosophy within Smith &

Nephew is that Sustainable Development considerations should be part of everybody's job, which will mean more training.

There is already a demand for business to show clearly that it is behaving in a responsible manner to the important issues of Sustainable Development. Some guidance for companies has already been provided by National Association of Pension Funds (NAPF), Association of British Insurers (ABI) etc. However, this demand will intensify in the near future given the lack of confidence in some companies' behaviour, particularly in the USA.

1.2 Sustainable development in higher education

Universities also need to recognise within the educational requirements the growing demands from industry and also to provide the necessary educational material for their students. Engineering and design education can contribute to sustainable development within industry by providing graduates with the relevant tools. Engineering designers educated at university become key people within the product, process, and packaging disciplines within the business. The designer decides what materials to use, the manufacturing process/equipment selection, and decides how the customer should use and dispose of his product at the end of its life. So the designer needs to take into consideration the environmental, social and economic implications of design proposals.

1.3 Educational and industrial challenges

- Business and universities need to understand each others' requirements, limitations and educational training needs

- Sustainable development is a very complex subject covering many areas of expertise of which the designer/engineer has limited knowledge

-

- Difficult decisions have to be made on complex issues and often there is not one correct solution

-

- Individual academic staff need to respond to the explicit drivers

- Seen sometimes as yet another initiative to be included in the curriculum or daily workload

- The key is to integrate sustainable development via awareness, knowledge and experience into all our jobs

In this context 'education' is defined as awareness and understanding of the principles and concepts of sustainable development and 'training' follows in providing a practical application and the competence to use appropriate tools and guidance. It is vital that educational facilitators and industry provide a partnership to embrace the educational and training aspects of sustainable development.

2. SUSTAINABLE DEVELOPMENT EDUCATION FOR ENGINEERING DESIGN STUDENTS

2.1 Background

The UK Royal Academy of Engineering (8) has introduced in 1997 the idea of appointing Visiting Professors at various UK Universities to provide lecture material/case studies to promote Sustainable Development in Engineering Design for engineering students. The Visiting Professors are from business and industry and provide a clear synergy between Industrial Good Practice and Education. This paper describes in more detail the work at Bournemouth University (Professor Mark Hadfield) and the Visiting Professor, George Howarth, Environmental Affairs Director, from Smith & Nephew- a major International Medical Devices manufacturer.

The approach taken is as follows:

(a) Build up from three simple but fundamental concepts
- Sustainable Development assessment needs to evaluate the three **Environmental, Social and Economic aspects** of the design/project
- The assessment should be holistic, i.e. from raw material selection through manufacture and distribution to use and then final disposal at the End of Life, i.e. **Product Life Cycle**
- The views/attitudes of **'stakeholders'** who have an interest in the business/organisation should be considered in any new design/project

(b) Although the major "customers" are design students, the training material should be suitable for other functions within business – purchasing, technical, marketing etc.

(c) The material will initially concentrate on concepts/ideas but it should be able to provide detailed information if required by the designer

(d) Need to provide simple practical tools the student can use

(e) Ideally should be web-based for ease of use, updating etc.

(f) The engineering designer should have control and ownership during project based learning and therefore an interactive environment is essential

2.2. Learning environment

Awareness, understanding and integration of environmental issues within the design process are the underlying issues of this project. A holistic philosophy to design is the ethos of the design courses at Bournemouth University and thus sustainable development is a key issue. Specifying products that meet the needs of the present generation, without compromising the ability of future generations to meet their own needs, is the challenge for designers. Sustainable development needs to address issues that include economic growth and employment, social progress for all, protection of the environment and prudent use of natural resources. This is a complex and multidisciplinary subject and thus modern information technology is embraced to facilitate and communicate effective industrially relevant case studies.

The purpose of this project is to produce case study teaching material to relate sustainable development issues to undergraduate students within the School of Design, Engineering and Computing (DEC) at Bournemouth University. Most of this information was obtained from Smith & Nephew and more will follow as described later. It is intended any teaching material produced at Bournemouth University and within Smith & Nephew will be exchanged. The two organisations do have different educational perspectives.

Bournemouth University perspective
There are two education processes – traditional lectures/seminars and active learning using the Internet as a vehicle. The lectures cover the issues and concerns related to the elements of sustainable development and the constraints on a designer plus details of design tools using check lists etc.

The Bournemouth Sustainable Product Development website (9) first provides an introduction into Sustainable Development and Product Life Cycle and then at the "home page" can take the student into practical case studies – one is a 'Life Cycle Assessment', another on 'Design for Waste' and one more on Sustainable Product Development Assessment is to follow. In addition the student can view the Lecture Notes, Glossary of Terms and Sources of Information. For the student who is familiar with Sustainable Development and LCA, he can go to a "Tools" section to use the LCA programme for examples on their new design (see Fig. 1). Some pedagogical components principles (10) applied to website structure is as follows:

- Variety (learners' attention is engaged with a range of stimuli – e.g. photographic illustrations, dynamic schematic/block diagrams, hyper links with other web sites etc. encouraged)
- Action (learner is doing things e.g. student to enter details, questionnaire/survey of environmental awareness with instant feedback, etc.)
- Application (learner should be encouraged to apply learning in another context e.g. hyper-link to personal feedback questions and illustrated examples within the study)
- Interaction (learner can change/comment on content – changes to materials, process routes, waste strategy, re-cycle strategy to be selected by the student and the effect on environmental impacts to be illustrated. Students will be encouraged to prioritise environmental impacts and show different design outcomes)
- Feedback (learner reflects on what they have done or understood – including direct e-mail link to tutor for questions, frequently asked questions available within each case-study, activity of each case-study monitored and presented, instant feedback within the assessment module, discussion forum for final year project students)

Smith & Nephew plc perspective
There has been for over ten years an Environmental Management System in operation at all manufacturing sites, which concentrated on the process environmental impacts. There is a need to make all staff far more aware of the social and economic aspects besides the environmental impacts. In addition the company has made the commitment to introduce Sustainable Product Development Assessment. Smith & Nephew wanted not only to educate the product designer on Sustainable Product Development, but also other technical staff and at least make staff in purchasing, marketing etc. aware of the issues and business implications.

The company started with the Sustainable Development concepts, issues and definitions but, for example, made the definition more related to their own operations. It was important to first define the general structure of the Sustainable Development Programme by defining the elements to consider and the possible practical tools for use by staff (Fig. 2). The Product Life Cycle is key and it was important that each function involved in the various parts of the cycle understood the issues, concerns, drivers etc. and to better understand their possible contributions.

Having made staff aware of the issues etc. it is important to provide some practical design tools, for example hazardous materials list, process impact form (see Fig. 3).

Following a review of this proposed training scheme, it was agreed by the product designers to concentrate on raw materials, as this is the key to any project. It would be very beneficial if inappropriate materials could be eliminated at a very early stage in the product development programme.

This Smith & Nephew materials system is based on the Bournemouth University Sustainable Product Development website, but with the addition of a check list of aspects to consider when selecting a material. Links/references to further sources of information will be provided. This will be a website and it is intended that once completed this will be added to the Bournemouth website as part of the "Tools" available to the students.

The second training project was related to the introduction of the Sustainable Product Development Assessment model for an **existing** product.(11) Now the designer was not included; instead purchasing, manufacturing, technical and marketing staff. In addition a much more broad brush approach was needed. However, this time the Product Life Cycle included social and economic aspects. Typical issues/concerns to consider in identifying opportunities and risks are given in Fig. 4.

Initially, given the above awareness/training each functional team was asked to prepare their own views as to what were the Sustainable Development concerns and issues. The project leaders did not want to force a solution, however they needed to make sure all aspects were covered and a common understanding agreed, so a joint discussion document was used. This resulted in a list of possible opportunities for adding value and some potential risks within the operation.

A pilot desk study has been completed on an existing Smith & Nephew product and this work has formed part of the lecture and case study material at Bournemouth University.

This simple training scheme on Sustainable Development, which can be applied to many of the functions within business, will provide benefits to the business in identifying added value opportunities and potential risks.

3. FUTURE WORK

- **Bournemouth:** The website has a new case study on 'Design for Waste' which is in form of electronic book and is source of guidance and practical advice on waste minimisation and prevention for the product designer. There will be a case study on Sustainable Product Development Assessment Model. Next year the

'Materials' tool will be added, but the major effort will be in consolidating this initiative within the whole of the Design, Engineering and Computing school.

- **Smith & Nephew:** The training will be extended into our overseas operations and further Sustainable Product Development assessment will be made of existing products. There are two key areas of concentration in future – people – their awareness and performance plus New Products sustainable development evaluation.

REFERENCES:

(1) UK Government Sustainable Development strategy "A Better Quality of Life" – May 1999. http://www.sustainable-development.gov.uk

(2) European Commission: "6th Environmental Action Programme, Environment 2010" - COM (2001)31

(3) Commission of the European Communities: Green Paper –"Promoting a European Framework for Corporate Social Responsibility" — Brussels 18.7.2001 – COM (2001) 366 final

(4) Commission of the European Communities – "A Sustainable Europe for a Better World: A European Union Strategy for Sustainable Development" – Brussels 15.5.2001 - COM (2001) 264 Final

(5) Department for Environment, Food and Rural Affairs News Release Feb 2003. http://www.defra.gov.uk

(6) Global Reporting Initiative (GRI) (http://www.globalreporting.org/)

(7) Smith & Nephew Sustainability Report, www.smith-nephew.com

(8) Royal Academy of Engineering, www.raeng.org.uk

(9) Bournemouth University SPD website – www.spd.bournemouth.ac.uk

(10) Hutchings. M, Lewarne. S, Norman. K, Garland. N, Hadfield. M, Howarth. G. "Educational challenges of web-based studies in sustainable development Design and Manufacturing for Sustainable Development", 1st International Conference, Liverpool, June 2002, PEP Publications, ISBN 1 86058 396 2

(11) Howarth. G, Hadfield. M. "An assessment model for sustainable product development", International Conference: TSPD7 Managing Sustainable Products, London, 28-29th October 2002, pp. 147-157. ISBN 0-9543950-0-X

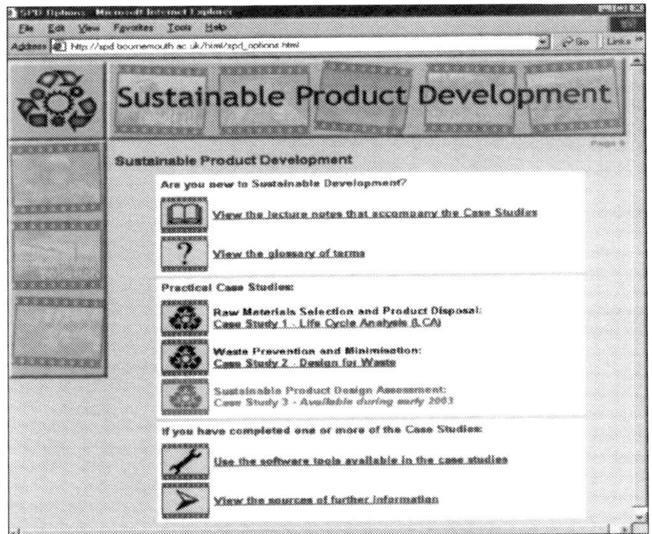

Figure 1. Bournemouth SPD website

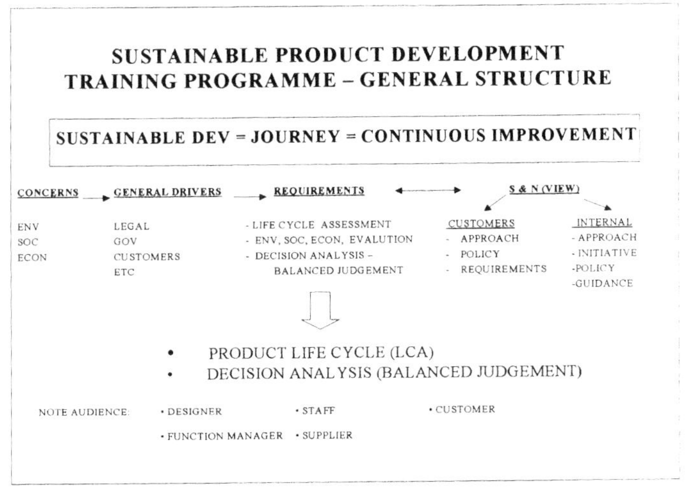

Figure 2. General structure for training programme

SUSTAINABLE PRODUCT DEVELOPMENT TRAINING PROGRAMME – DESIGNER TOOLS

RAW MATERIALS*	MANUFACTURE*	DISTRIBUTION*	USE*	DISPOSAL*
• LIST OF HAZARDOUS SUBSTANCES RED/BLACK/GREY •LIST OF ENV. DAMAGE • LIST OF S & N MATERIAL CONCERNS •ENV LEGISLATION WASTE DIR HAZ WASTE IPPC ETC	•PROCESS IMPACT FORM • PERMITS LIST	• TRANSPORTATION SYSTEM IMPACTS & REQUIREMENTS • LEGAL REQ. • SERVICES PROVIDED	• LABELLING REQUIREMNTS • DISPOSAL INFORMATION •TECHNIQUES - EXT. LIFE - SIMPLICITY - MIN. USAGE - COMMON MATRIALS	• DESIGN FOR REFUSE RECYCLING DISASSEMBLY • LEGAL PPWD WEEE • DISPOSAL METHOD - LANDFILL - INCINERATION - COMPOST

NOTE: DEPTH OF INFORMATION PROVIDED WILL VARY FOR EACH NPD STAGE

Figure 3. Designer tools

OBJECTIVE – DEFINE RISKS & OPPORTUNITIES

SUPPLY CHAIN / VALUE CHAIN / PRODUCT LIFE CYCLE

	SOURCE	SUPPLY & TRANSPORT	DESIGN	MANUFACTURE	DISTRIUBTION	USE/ CUSTOMERS	DISPOSAL
ENVIRONMENTAL	•Min Environmental Impact •Risks identified	•Fuel consumption •Less emissions •Air flights	•Min. material •No hazardous materials •Disassembly •Recyclable	•Min. Energy •Min. Waste •Min. Emission / Discharges / Special Waste •Nuisances Reduced	•Efficient fuel consumption & less emissions	•Less energy, waste emissions •Less packaging	•No hazardous material •No heavy metals •Recyclable
SOCIAL	•Health & Safety •Child Labour •Human Rights •Community impact	•Local Community -Frequency -Airport	•Animal Tests	•No nuisances •Local impacts reduced	•Fewer journeys •Utilisation high	•Safe for patient & healthcare professional •More efficient time & resources •Improved services	•Safe for incineration & landfill
ECONOMIC	Costs	•Costs	•Costs	•Costs •Environmental & social damage	•Costs •Environmental & social damage	•Health Economics •Community / individual / cost & benefits	•Collection & Treatment costs

Figure 4. Smith & Nephew sustainable product development model

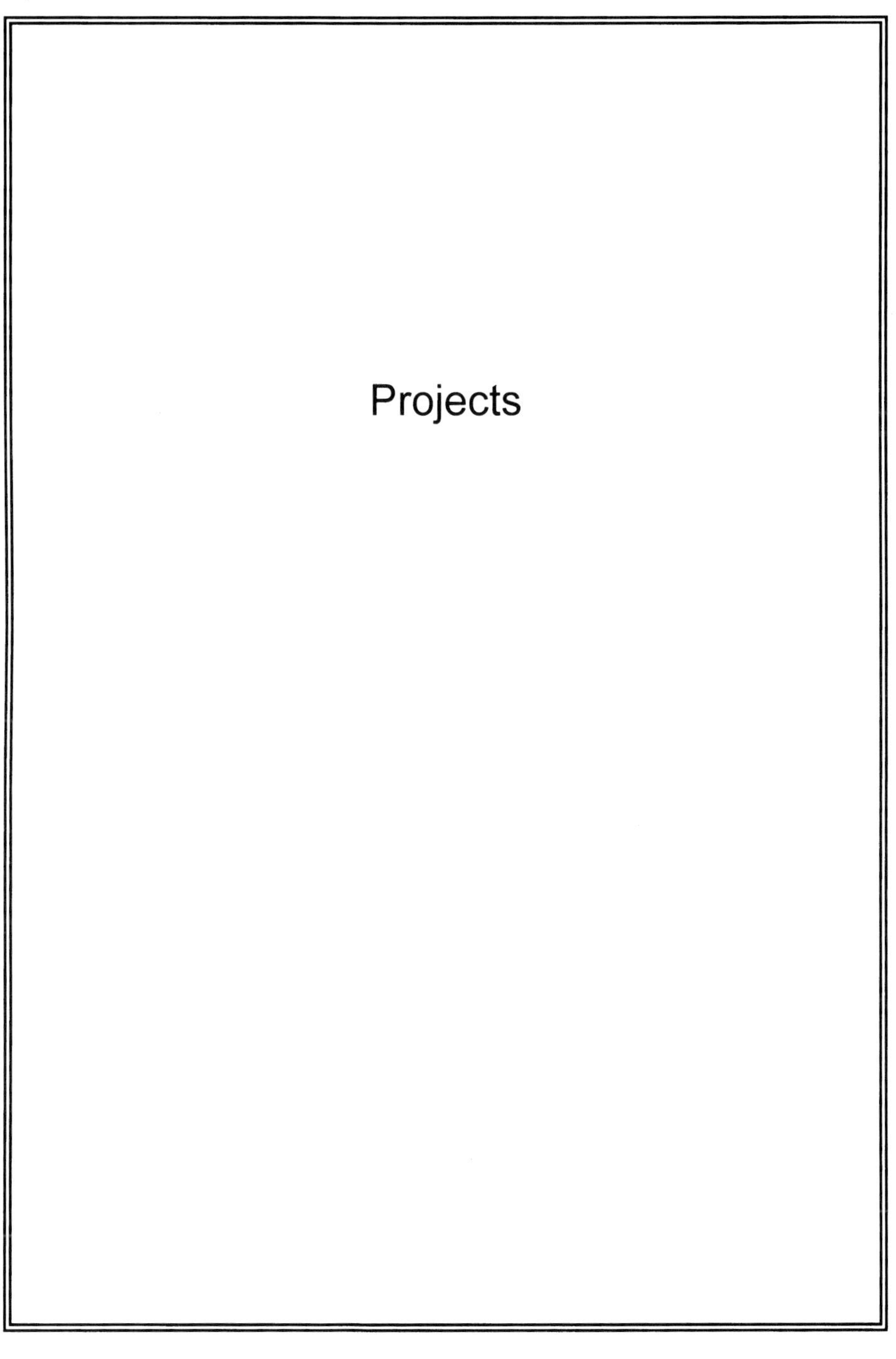

Projects

Sustainability – a design exercise?

M A C EVATT
School of Engineering, Coventry University, UK

ABSTRACT

This paper relates the development of a second level module of Design Studies offered to students following a BSc Engineering and Business Management Degree Course at Coventry University. Students were asked to explore problems of environmental concern and the designer's part in the design of environmentally friendly artefacts.

A parallel study administered a short questionnaire to a number of groups of students to assess their understanding of commonly used phrases such as 'Green Design', 'Sustainable Design' and 'Design for the Environment'. Results of this exercise have also informed the paper which mainly focuses on how students following the relevant module were able to develop a design tool which could meet the requirements of the brief given.

1 INTRODUCTION

Design is a mandatory part of the course diet for students following the BSc Engineering and Business Management Degree Course at Coventry. Although it is recognised that these students will not become professional designers they do need to know what designers do, the issues that designers have to address and how design is processed and managed.

Environmental impact is considered to be a key area of design that should be investigated by students on the course rather than pursuing more traditional topics such as design tools and machine elements.

From previous unpublished work it was realised that students, and others, lack basic awareness of world problems that impinge on the design activity. It therefore seemed appropriate to set those following this programme the task of discovering for themselves the problems of environmental concern and the designer's part in the development of environmentally friendly artefacts. It was felt to be important that the students carried out their own investigation so as to actively build their understanding of these issues rather than merely attempting to tell them of the issues within the passive context of a lecture.

These students will normally have completed a first level module which looks at design process methodologies and creative thinking techniques and have completed several design assignments. The new second level module was therefore developed to contain three main elements:

- An opportunity for students to revise, reinforce and extend their first level work on design and idea generation.
- An information retrieval and costing exercise on horticultural equipment such as mowers, cultivators and strimmers.
- A research-based element aimed at developing an understanding of the issues involved in carrying out eco-friendly design and producing a list of factors to be taken into consideration when carrying out such design work which they had generated for themselves (rather than taking as a given from the work of others).

This list providing an additional design tool or template for the completion of the module which requires a redesign of a (horticultural) product or their choice. The module is continuously assessed rather than subject to formal examination.

This paper seeks to document the students' progress in developing their eco-friendly design template as well as reporting the outcomes of their re-designs. It will also report on the findings of the questionnaire study and comment upon the success of the activity and discuss any modifications deemed necessary for future use.

2 THE ASSIGNMENT

The assignment description handout is reproduced below: -

Level 2 Design Module Final Design Assignment - Design for the Environment

Stage ONE

1. Students in Groups of no more than FOUR will investigate what is meant by the following: - Green Design, Environmental Design and Design for Sustainability and derive suitable definitions. **15%**

2. Students will research using questionnaires etc. the opinions of the student body and the general public to establish their understanding of the issues, what they are doing about it, whether they would pay for the benefits. **15%**

3. Each group from the above and perhaps additional research will derive a set of rules (perhaps 10 bullet points will be sufficient) to enable a designer to design an environmentally friendly and sustainable product. **20%**

The above research, evaluation & recommendations will form the front end of the final report.

Stage TWO

The students in the same groups will now undertake the re-design of a 'garden machine' of their choice taking due account of the above rules together with the use of the Design Methods and Creativity Tools highlighted earlier in the module.

Design Review **10%**

Oral Presentation of Final Design 10%

Report Submission 30%

Students are reminded that this assignment constitutes 60% of the module mark

3 ANTICIPATED OUTCOMES

It was expected that the students would develop a check-list similar that developed by Burall (1) who suggested the following:-

The environmentally-aware designer should aim to: -

1. Increase efficiency in the use of materials, energy and other resources.
2. Minimise damage or pollution from chosen materials.
3. Reduce to a minimum any long-term harm to the environment caused by the use of the product.
4. Ensure that the planned life of the product is the most appropriate in environmental terms, and that the product functions efficiently for its full life.
5. Take full account of the effects of the end-of-life disposal of the product.
6. Ensure that packaging, instructions and overall appearance of the product encourage efficient and environment-friendly use.
7. Minimise nuisances such as noise and smell.
8. Analyse and minimise potential safety hazards.

This checklist would then be used as a design tool to re-design their chosen artefact.

4 THE PARALLEL QUESTIONNAIRE

The findings from a parallel exercise of asking a group of mature students and a group of first year product design students to define Sustainable Design, Environmental Design and Green Design were quite interesting.

There was an indication that some thought that **environmental design** was more to do with making the built environment better and that **sustainable design** was to do with products being made to last longer. Whereas others thought that **green design** was merely a marketing catch-phrase.

However the majority of comments and definitions were close to the usual understanding of the terms. There were differences in the responses of the two groups with the 1st year students presenting a more focussed and complete picture whereas the mature students seemed to concentrate more on recycling, re-claimed materials and natural and organic products. Perhaps this latter comment is a reflection that levels of awareness have increased in the dozen or so years since Burall first penned his list.

5 OUTCOMES OF THE ASSIGNMENT

The questionnaire

The majority of the groups produced a questionnaire to facilitate the harvesting of public and student views and opinion.
A typical example is seen below: -

Sustainability Questionnaire

1. What do you understand by the word 'sustainability'?
2. Is sustainability of a product particularly important to you?
3. Do you purposely buy products that are sustainable and/or environmentally friendly?
4. Would you pay more for a product if it was sustainable?
5. What do you consider more important:
   A product with high quality, but low sustainability
   A product with a lower quality, but more sustainable?
6. Would you still buy a product if its manufacturing harmed the environment?
7. Could you name any companies that are renowned for product sustainability/environmentally friendly?
8. Do you know any specific legislations/laws on sustainable/environmental design?

Most groups commented that the information thus gained showed that the general public were ill-informed regarding environmental issues and furthermore were not particularly interested in them. It was commented that most people appeared to be more interested in value for money when purchasing a product rather than sound environmental practice. It also seemed apparent that people were quite prepared to adopt good recycling habits providing the local authorities facilitated this with suitable receptacles and collection thereof. Only a minority said that they would pay extra to have a 'greener' product – the only exceptions being where they could see an actual monetary advantage to themselves – lower gas or electricity bills for example.

The Definitions

Just as in the parallel survey reported in section 4 above some misunderstandings were found although for the most part quite adequate definitions were proposed. Typical responses were as follows: -

Student Group One
Green Design explained as products or services that are designed using as many natural resources and processes as possible. This includes minimal usage of man-made materials

Sustainable Design describes thorough design that can stay useable and relevant for a long period of time and be mended/maintained without the need for full product replacement. The abolition of 'built-in obsolescence'

Environmental Design is related to the interaction between extraction, production process, usage and finally disposal or a product that causes minimal damage to the environment.

Student Group Two

Sustainable Design is described as "development that meets the needs of the present without compromising the ability of those in the future to meet their own needs".

Environmental Design focuses on the relationship between people and the built environment. It has the potential to improve the quality of life for its users.

Green Design is the attempt to make new products and processes more environmentally benign by making changes in the design phase.

It was noticed that some of the anomalies were in part due the students' lack of expertise in crafting the appropriate sentence rather than any lack of understanding. This was evident when they were questioned about their responses.

The Check-lists

This part of the assignment was also carried out in a competent manner with most groups producing bullet points aligning quite closely to Burall's version. This is exemplified by the two representative lists shown below: -

List A

1. Minimal material usage
2. Basic, non composite materials used where possible
3. Materials must be able to be separated at End Of Life
4. Components must be able to be separated at End Of Life
5. Manufacture has minimal effects on surroundings
6. Standardised components
7. Replaceable components
8. Bio degradable non-recyclable parts
9. Long lasting recyclable parts
10. Power efficiency

List B

1. Select and use materials with reduced environmental impact.
2. Design product to be easily disassembled for the process of recycling.
3. Aim to reduce the amount of material used, and the products weight.
4. In addition to making disassembly easier, designers should strive to cut assembly time.
5. Packaging should be either: re-usable, biodegradable or recyclable.
6. Reduce and ultimately remove pollutants.
7. Reduce electrical and domestic products' energy consumption.
8. Design products to utilise and accommodate renewable 'green' energy sources.
9. Monitor legal regulations (from local to international levels) regarding the environment so that they can be anticipated.
10. Cooperate with and liaise with groups and organisations concerned with the protection of the environment for feedback, advice and improved relations and promotion of programmes and agreements.

It was interesting to note that recylability and bio-degradability featured in most lists whereas they are not in Burall's original and that some of the points are not really design pointers that can be used to further environmentally friendly design.

The Re-design of a 'Garden' Product

In some ways this part of the assignment was quite disappointing but maybe it was just that the author's expectations were higher than these students, not seeking to be "design specialists", were ever going to deliver. It is nearly always the case that when students become slaves to the design process then the quality of the design outcomes suffers. However the majority of the groups made a reasonable effort to design a product to the set of rules that they had generated. Some products were more overtly 'greener' than others such as battery powered lawn-mowers with solar powered charging facilities whereas others concentrated more on the use of sustainable materials and processes.

The downside is that because to the vast amount of relevant material out there on the web some students homed in on this and used it as a basis for their 10 factors. On the other hand some groups seemed determined to plough their own furrow and re-invent the wheel or at least propose a list of factors – which is more worthy? This is an increasingly difficult area to police and assess effectively (a possible subject for a future paper?).

However in a feedback session the students generally agreed that they had gained a lot of knowledge about environmental problems by having to do research into the topic themselves. They believe that they now have greater understanding of the issues, ideas and conflicting views associated with the whole notion of eco-friendly approaches to design than they would have had by receiving the information in lecture mode.

6 CHANGES FOR THE FUTURE

Little change is anticipated for the next running of the module apart from the introduction of guidance on the use and referencing of materials gleaned from the internet. It is also intended that a rather longer period of time be given for the final design activity in an attempt to improve the quality of the outcomes.

7 CONCLUDING REMARKS

It is considered that the original objectives of the new module have been met in that the activities were appropriate for the student cohort and that the research skills gained by the students can be applied in other areas and that some real learning had taken place regarding the world's environmental problems.

8 REFERENCE

Burall P. "*Green Design*" The Design Council 1991 page 16

Cabin and passenger environment design for the Airbus A380 – a case study for education

R D MORRIS
School of Engineering, University of Brighton, UK
A L THOMAS and **P R N CHILDS**
Department of Engineering, University of Sussex, UK

ABSTRACT

This paper describes the experiences of using the Airbus A380 as a platform for teaching design in a second year B.Sc. Product Design course. The study uses observation and student experience to assess the success of the exercise, and seeks to inform the current debate on "Problem Based Learning" in design tuition. The emphasis is on the way the project-aims, timeliness and real-world relevance and its extended period of study, long enough to explore and integrate alternative ideas, rather than its size, has contributed to a successful learning exercise that has engaged full student commitment over the one year duration of the project.

1. INTRODUCTION

The potential aims in using a project-based problem for teaching "design" were manifold. The key objectives addressed in this course were to:

i) provide students with the experience of the design process,
ii) develop student confidence in tackling a design problem,
iii) provide a platform for students to learn and apply design related techniques,
iv) provide opportunities for students to improve group working abilities.

The traditional emphasis to education within engineering is on the structured study of key analytical techniques. Available time and resources often limit the reality of the exercises and solutions that can be explored. The aim is to provide a problem solving "tool-kit", so the student is capable of handling a range of problems that might be met in a professional career, often in as efficient an educational manner as possible. However, while the relevance of the techniques being presented using this approach may be understood by the lecturer, where the student does not have the experience to place them in a working context, much of this work has to be taken on trust, anyway in the short term, and cumulatively this can make it easy for a student to lose direction, focus and motivation.

Over the past two decades, "Problem Based Learning" has been introduced as a new approach to subject areas like engineering and medicine which demand a high level of technical knowledge: challenging the idea that a corresponding high level of structured teaching is necessary "*before* realistic work can be undertaken". This "new" approach *starts* by addressing an open-ended real-life problem rather than smaller simplified exercises chosen to make particular analytical techniques tractable. Although such an unbounded task quickly generates questions, which need the analytical techniques provided by the traditional courses to answer them fully, they arise in a working context that gives them relevance and a relative importance within an overall understandable framework.

This approach has grown in popularity since its ability to relate knowledge learned in class, to real world problems in health education, was high-lighted by [1]; see also [2]. It has subsequently been shown to be relevant in a variety of educational contexts, for example, the teaching of English Literature in Manchester University, and Entrepreneurship at the University of Brighton. However, it is already a well-established approach to teaching and learning in professions such as Architecture, Civic Design and City and Regional Planning. In a sense it can be regarded as a replacement for some aspects of engineering education traditionally covered by apprenticeship schemes. Where students are training for the profession of Product Design and need to manage the wide range of knowledge required to cover the technical, social and economic issues involved, its adoption at the heart of the curriculum seemed totally appropriate.

2. THE AIRBUS A380

In this example of "problem based learning" the challenge of designing the cabin of the Airbus A380 was given to eleven second-year students. The Airbus A380 shown in Figure 1 is a multi-billion Euro activity involving design and manufacture across the world. Its specification was defined in close collaboration with major airlines, airports and airworthiness authorities, resulting in a fixed airframe specification. However, the cabin interior was left flexible in the way shown in Figure 2, to suit the different airline's perception of their customers' requirements.

Figure 1 The Airbus A380 (Courtesy of Airbus)

Cabin design involves the consideration of many issues: emotional aspects such as customers' perception of aesthetics, personal space, safety and service efficiency; physical aspects such as vibration and sound transmission, heating and air-conditioning, odour control and ventilation, and artificial and natural lighting; spatial aspects such as circulation and access, seating arrangements and the ergonomics of customers' sitting, sleeping and storage requirements; and constructional aspects such as strength balanced against the need for lightness, and the requirements for flexible layout and maintenance. However, the cabin design is only one concern facing an airline business, which must also cover marketing, booking, check-in and departure facilities, and baggage management and retrieval facilities.

The A380 cabin design was selected as a project for the course, while it was still in its design and development stage, the first commercial flights being scheduled for 2006. This gave the exercise an excitement and immediacy resulting from students exploring the same issues being addressed by large multi-disciplinary teams, at the same time that important design and commercial decisions still needed to be made. The challenge facing the students, in its broadest extent, therefore covered the complexity of a multidisciplinary, international design activity. The difficulties the size of this task presented, made clear the need for effective project management techniques from the beginning of the exercise.

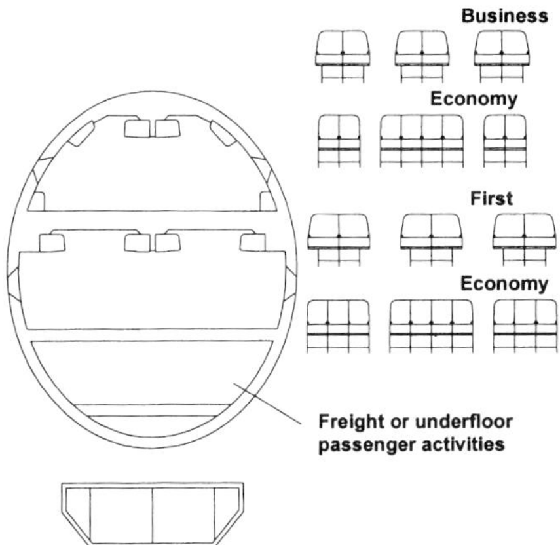

Figure 2 A380 Cabin layout

3. DELIVERING THE A380 CHALLENGE

While a prerequisite for "problem based learning" is a suitable problem, this approach can also be characterised by the flexibility and the diversity of the different ways in which a problem may be handled. In this case three issues were directly addressed: *group working*, *problem structuring* and *problem framing*.

3.1 Group working

It is an on going educational debate as to whether "problem based learning" should or should not be conducted by groups or by individuals. Where group work is undertaken then group sizes of four or five people are often cited as optimal teams. In this case with eleven students, and with the danger of losing one or two team members, two teams of five and six members gave the only practical arrangement, if each team was to tackle the whole problem.

3.2 Problem structuring

The structuring of the project in terms defined in the syllabus was inflexible. The project ran over three terms in the second year of a B.Sc. in Product Design at the University of Sussex. The project was assessed by means of course work and counted for 30% of the marks for the year. Although the "timing" and "deliverables" of the project work were potentially flexible, it was decided, to provide a semi-rigid framework by providing milestones which student would have to meet. These were presentations of 1) an initial market and technical-research study, 2) a broad concept design study exploring the overall cabin treatment, 3) followed by a detail cabin layout study, then 4) a detailed seat study, finishing with 5) a full scale mock-up of the sitting arrangements in a section of the cabin.

Each of these studies involved finding the most suitable modelling techniques for the design decision-making that was appropriate to the tasks being addressed. These included: schematic, diagrammatic, presentation and technical drawings; scaled, Figures 3 and 4, and full size physical models, Figure 5; and computer and related mathematical models, using software. The original layout of the stairs between the A380's decks, initially designed in plan drawings had to be changed when the $1/10^{th}$ scale models shown in Figure 3 were built and the three dimensional relationships involved could be perceived and manipulated directly. Similarly the full sized model shown in Figure 5 allowed design judgements to be made about human-scale effects using rapid construction techniques and relatively crude materials. Although the aim was to solve the design problem, the success of the exercise to design cabin layouts was decided by the learning objectives defined in an assessment framework, which sought to ensure that a professional design process was being developed.

Figure 3 Model making

Figure 4 Quick layout-testing using scaled models of chairs

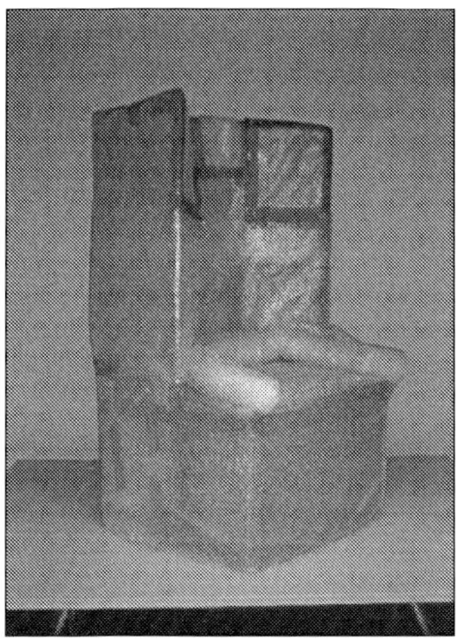

Figure 5 Rapid seat modelling to judge human scale

3.3 Problem framing

The overall assessment framework was provided by the system based analysis summarised in Figure 6. The "context" to the cabin design was defined by the economic and market analysis studied in the first term. This work set out target seat counts, which were related to ticket pricing, levels of comfort and the nature of the service. Airframe size, range and speed and therefore likely routes, provided some of the technological constraints.

Context

External System Requirements

Design Problem Area

Technological Constraints

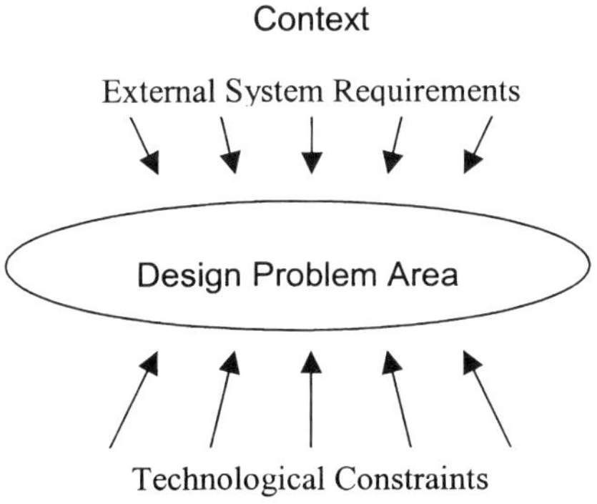

Figure 6 Design system analysis

The decisions made about cabin layout provided the "context" for the more detailed study of the aircraft seat design. Enumerating all the activities that would be carried out in the cabin and the functional demands they made on the seats, set up the context, which established the specification and design requirements for "a seat", studied in the second term. This "context" also included: space constraints for ergonomic design, safety requirements, aesthetic issues, and environmental controls for heating, lighting and ventilation. Constructional requirements for adjustable layout, lightness of build and ease of maintenance, represented some of the technological constraints that applied. Some of these issues could be analysed by the students based on their existing knowledge, some required knowledge they did not have. Acquiring this missing knowledge or information could be tackled in two ways, the first requiring further personal study and research to provide the necessary answers, the second commissioning expert help. The latter case would be handled in real life by specialist consultants, or from cooperating specialist teams. The subdivision of the overall work within the student teams, allowed the task of cooperation between specialist groups to be explored, because the work load was partitioned and distributed to team members to work up and then report back and integrate into a working whole.

Two inputs from "expert consultants" were arranged from faculty engineers, the first offering structural advice the second giving the basic relationships which would control the layout of the ventilation and air-conditioning ducting. In both cases the support was only offered when demanded by the students' study needs. An interesting outcome from this arrangement was the totally different perception of problems as seen by the students and that initially assumed by the experts. The experts' initial response to the students' request for help was that there was not time to teach them how to solve their design problem properly, whereas, all that the students wanted, in the first instance, were ways of bounding the solution within viable size ranges and into practical layouts, either by using simplified formulae or reasonable rules of thumb. The need to explore and service the communication needs set up by the interface between different specialist engineering design groups, as part of this courses content, was clearly highlighted by this experience.

Once independent partial studies were complete there followed the task of bringing together all the requirements into a single integrated design solution. This was where this approach diverged from merely analytical engineering teaching methods. Solutions for sub systems and sub-assemblies had to be compatible with each other. This meant exploring alternatives, both as simple and complex solutions, to bring together a working design, which usually involved compromises, and priority choices being made. Where the two teams made different design choices, it provided very useful discussions for all concerned.

4. OUTCOMES

In addition to tutor observation, students were asked to provide qualitative and quantitative feedback to help determine if the A380 challenge met the teaching aims, Table 1. The outcomes of this are summarised below.

i) *To provide students with the experience of the design process*

Most students benefited in some way from the exercise, though differences were noted depending on the background and previous experience of the particular student.

Typically:-

"Designing a product intended for a particular market I always thought took more effort than one would first think. I didn't realise however, until completing this course [module] just how much is done behind the scenes for the generation of concepts and final design development. The course [module] has provided me with an incredibly detailed insight into the stages a product goes through before it is available for purchase, and in an engaging and captivating manner". [Chris Davies]

ii) *To develop student confidence in tackling a design problem*

At the time students undertook customer related research, the airline industry was split between a high value, business approach (British Airways, Virgin) and a low cost approach (EasyJet, Go). Industry and marketing experts would have predicted the decline of the high cost arrangement but neither student group had the expertise or the courage to adopt this approach. The approaches adopted by the students were either a mix of low cost and high value seating, or a mid range option with low cost seats with high value extras. During the course of the project, the high value airlines began to publish poor trading figures compared to the profitable low cost airlines and even British Airways began to remove high value seats on some of its routes. Students were able to reflect that they could have been bolder with their own design decisions based on the facts that were originally available. Students were also able to reflect on the diametrically opposed design concepts that emerged. One group developed a strongly branded, bold concept whilst the other developed one that was neutral and flexible.

"This has been a valuable experience and my favourite course [module] so far." [Lawrence Weyman-Jones]

iii) *To provide a platform for students to learn and apply design related techniques*

55% of students expressed the value of learning to communicate through presentations, 44% singled out creativity and 44% also singled out other design techniques. Typically:-

"I think this project has been a great experience to me. It makes you learn so much, so many different design skills, stages – using programs, working in teams, getting things organised. It is really the most enjoyable project in my degree at the moment." [Annie Li]

iv) *To provide opportunities for students to improve group working abilities*

The value of the group working experience was acknowledged by 88% of the students. "We met in the library three times a week, We even met up four times on Sundays which must show the dedication of the group". [Chris Davies].

Table 1. Student feedback

Learning experience issue raised by students	Percentage of students raising the issue
Group dynamic	88
Presentation	55
Creativity	44
Design techniques	44
Hands on	33
Customer needs	22
Technical research	11

5. CONCLUSIONS

Discrete independent design studies are not as demanding as finding solutions in studies targeting an overall design objective, because in the second case the partial results have to be compatible with each other. This project illustrated this in many ways. Although the importance of providing an overall goal was recognised, the size of this project made it unrealistic to expect all the design-exercise goals to be reached. By providing a structured approach with milestones for presentations, it was possible to satisfy the four key aims to develop design experience, self-confidence, design support skills, and team working strategies. The subject was chosen to ensure multi-faceted, multi-disciplinary design issues had to be addressed, so techniques to handle them could be explored. As a first run it was felt that the course had been successful in meeting its teaching objectives.

Success has to be placed in context with the degree programme that surrounds the project. For example, if students had not experienced group working before, they might naturally place a higher emphasis on the experience of group working gained in the project, than otherwise. It did seem however, that students enjoyed and valued the experience, where they were able to support and be supported by colleagues in mutually beneficial ways through a highly demanding exercise. One result was that they were able to recognise the merits of a concurrent engineering approach as they explored the many different inter dependent aspects of a complex design task. There is also the question of how the course was delivered. Was the project well conceived as a teaching experience, or have we been lucky? It is hoped that the degree of tutor preparation and expertise indicates more than a lucky outcome, however only experience gained over a range of projects, over time, will answer this question. There was a conscious effort to keep the project successful by making the design experience as real as possible. This was achieved by making the student teams compete against each other in their presentations, and by continually updating the project with live, relevant and topical issues introduced by guest speakers, invited from the aircraft industry.

Finally the most interesting "emergent" benefit of the course arrangements was a consequence of the amount of time allocated to it. It was not the size of the exercise that was important, but because of its duration through the whole year, there was time to explore alternative concepts, to revisit ideas from various points of view, and most importantly time to "refactor" and "refine" design solutions as understanding of the problems deepened. However, the extent and nature of the project did mean that this process of reworking appeared less artificial, and was easier to manage without boredom and loss of motivation setting in. The experience gained working through the early stages in a design, when results seemed to be poor and disconnected, to the point when *"things started to gel and come together"*, is perhaps one of most important learning experiences needed for any design student! This experience is important in developing a student's confidence: to keep going through many false trials without abandoning an idea. The lesson that good design is more often the result of a systematic problem-space search, than simply inspiration, can be a hard one to learn. At the same time the ability to release an idea when it ultimately proves inappropriate especially after a lot of work is equally necessary. The selection process inherent in the contents of the film editor's cutting room floor is an important aspect of the design process. In this respect the importance of debriefing each student after each presentation, to help unravel their responses at each stage of the work and maintain courage and motivation until successful results started to give their own rewards and motivation, was vital.

REFERENCES

[1] Barrows, H.S. and Tamblyn, R.M. (1980). Problem-based learning. An approach to medical education. Springer Verlag.

[2] Savin-Baden, M. (2000) Problem-based learning in higher education: Untold stories. SRHE/Open University Press

Using small scale alternative energy equipment as a vehicle for sustainable development study

C McMAHON
Department of Mechanical Engineering, University of Bath, UK

ABSTRACT

Sustainable development, and in particular the need for alternative sources of energy, is one of the key challenges of the 21st century, and thus an important topic to include in the engineering design curriculum. This paper argues that design projects exploring the design of small-scale alternative energy devices are a good way to approach the study of sustainable development issues, because as well as dealing with such issues they allow other design topics such as design for manufacture and selection of product architectures to be addressed. The argument is illustrated with an overview of a series of projects dealing with wind, hydro and wave power devices, and a description of the design issues addressed in the projects. The paper concludes with the suggestion that the study can be extended to the engineering science and laboratory elements of the engineering curriculum.

1. INTRODUCTION

Sustainable development is increasingly seen as one of the key challenges of the 21st century. Of particular importance is the need to identify alternative energy resources in view of limited oil and gas supplies and of the likely effect of fossil fuel combustion on climate change (1)(2). Engineering designers will be central in meeting this challenge, and thus the incorporation of sustainable development issues into the engineering design curriculum is a very important current issue. The paper describes an approach to such incorporation through undergraduate projects that have explored a series of issues in small-scale alternative energy devices.

A number of alternative energy systems have been developed, including wind (3), wave (4), tide, hydro (5), geothermal and solar systems (6), and practical examples of all of these have been built. Most approaches to date are however of such a scale that it is impossible for students to build working devices, and difficult even for university research teams to resource. They are also extensively researched. However, small-scale systems have not been covered so thoroughly, are at a scale that models and even working prototypes are possible (although this has not been done so far), and they share many of the same issues as large-scale devices.

The projects that will be described here have all involved the exploration of small-scale alternative energy systems with a particular emphasis on design for large volume production - small-scale devices are of similar size in engineering terms to mass-produced products such as automobiles and white goods, and thus may share aspects of the manufacturing approaches. The reason for the emphasis is because if such devices are to be useful they will need to be used in large numbers and their price has to come down to economically justifiable levels, but also:

- It allows the economics of sustainability to be investigated – for example at what energy prices does investment in alternative energy devices become worthwhile, what are the wind/wave/tidal conditions (for example) for this to be the case, and what price does the equipment need to come down to for investment to be justified?
- It allows the study of other important design topics such as design for manufacture and assembly, parametric design and selection of product architecture to allow devices to match the greatest number of application conditions
- It allows issues of developing energy products for a market to be explored, and in particular investigation of what might be necessary to make alternative energy devices attractive to potential customers.

The work described here has formed the basis of individual undergraduate projects, but the same topics would make good subjects for group design projects, and the process of carrying out the projects generates a good deal of useful information that can inform courses on sustainable engineering and on engineering design.

In the next sections the projects that have been carried out will be described briefly, and then issues that have been emphasised in the projects such as design for manufacture and modularisation will be discussed in the following section.

2. PROJECTS CARRIED OUT

The work described here has been carried out at the Universities of Bristol and Bath over the past three years. A number of projects have been carried out that have investigated:

- Small-scale hydro-electric devices. These projects investigated the locations in which such devices could be applied, including rivers, streams and even drainage infrastructure, concentrating on the area around Bristol. They also addressed such issues as: what would be the most appropriate choice of impeller type for small-scale hydro plant? How can the devices be designed in a modular way to allow easy adaptation to different flow and head conditions?
- Small-scale wind turbines. Two projects are being carried out with small-scale wind devices as the target. The first is looking in particular at cost reduction and product design issues for wind turbines for domestic use. This study shows that the cost of installation of the turbine is one of the most important aspects. The project has considered whether existing structures such as street lamps and telegraph/electricity poles and pylons can be adapted to carry turbines, and how mass production techniques can be exploited in this. It is also looking at applying mass production techniques used in the manufacture of electrical hand tools in turbine construction.
- A second small-scale wind project explores whether a combined wind and solar device for a portable water pump and heater for use in remote areas is feasible (for example by troops on the move, campers and caravaners and so on). In particular this project addresses how to make a device very light and portable, and again issues of design for manufacture have been considered.

- Small scale wave energy devices. This project initially sought to identify whether equipment can be designed that easily adapts to existing structures such as breakwaters and harbour walls, but developed to consider offshore devices that could be produced in large numbers for use in "wave-energy farms". In that latter work the design issues explored have included the parametric design of oscillating water column turbines, and design for manufacture techniques for fabricated structures.
- For the wind and hydro-electric devices the sizes chosen were such that the devices would only generate a few hundred watts – sizes that put the equipment in the "micro" range. The size of the wave devices being considered is, by contrast, 10-20kW average output.

3. EXPLORATION OF DESIGN ISSUES

In this section the design issues that may be explored in small-scale alternative energy projects will be considered, with illustrations from the different projects that have been introduced.

3.1 Data collection for economic justification

Economic justification of alternative energy devices is based on a combination of factors: device characteristics and costs, environmental conditions and their temporal and geographical variation, energy prices, discount rates and other investment criteria. Device characteristics and costs are determined by the design process, but other factors need to be collected at the market investigation and specification formulation phase. Issues of particular importance in the selection and design of alternative energy devices include:

- The variation of wind velocity and direction with time, location and altitude. Location of wind energy devices is based on maps of mean wind velocity, and also on variation in wind velocity with height. The projects working in the wind energy area have obtained data for wind velocity at different geographical locations in the UK, and also of typical wind velocity variation with height for different ground conditions (e.g. for rural and built-up areas). The trade-off issues in the design of alternative energy devices are well illustrated by the trade-off in wind energy vs. tower height. As the wind turbine tower height increases the energy production increases significantly, but so does the cost of the tower.

- The flow rate and height of fall for the exploitation of hydro-power in streams and rivers. Students investigating micro-hydro devices worked with the Water Resources Officer of the Environment Agency to explore suitable sites on rivers and streams near Bristol, looking in particular at sites on a stream that fell some 200m in 25km. Flow data for the stream were obtained as long-term average flow (ADF) and flow rates reached 50% and 95% of the time. Heights of water fall at key locations such as weirs were also identified.

- Geographic and seasonal distribution of wave energy density, used for identifying the suitable locations and device design parameters for wave energy devices. Example values, for a site identified in the Bristol Channel, are shown in Figure 1.

3.2 Selection of design concept

For each broad type of alternative energy device there is a number of alternative conceptual design solutions that can be chosen. For example, for wind turbines the choice is between horizontal and vertical axis machines. For the latter there are various competing designs; for horizontal machines the main difference is the number of blades and the method for dealing with very high wind velocities. Even for the turbine tower there are several choices – between stayed or unstayed towers, tubular or lattice, and alternative methods of erection. For wave energy devices again a number of design principles can be applied, such as the Salter Duck, the OWEC buoy, the oscillating water column and so on. For hydro devices, the choice is between different types of impeller. Figure 2 shows the diagram used by the project students to choose impellers for different flow and head conditions in the micro-hydro project.

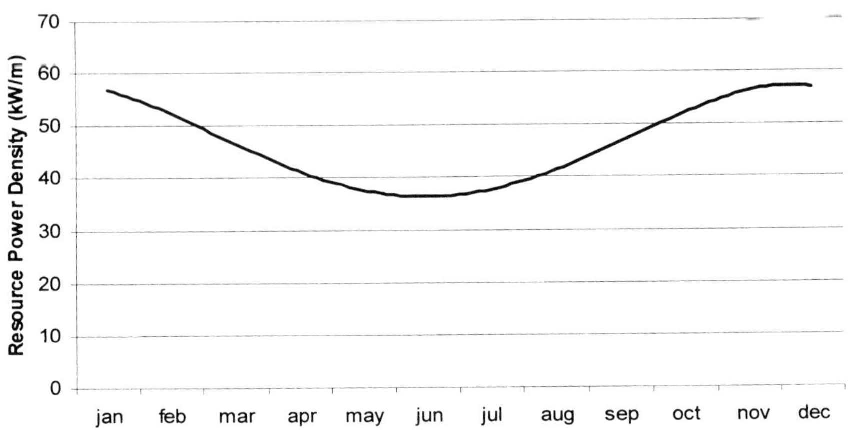

Figure 1: Seasonal wave energy variation at Lundy (51° N, 6° W)

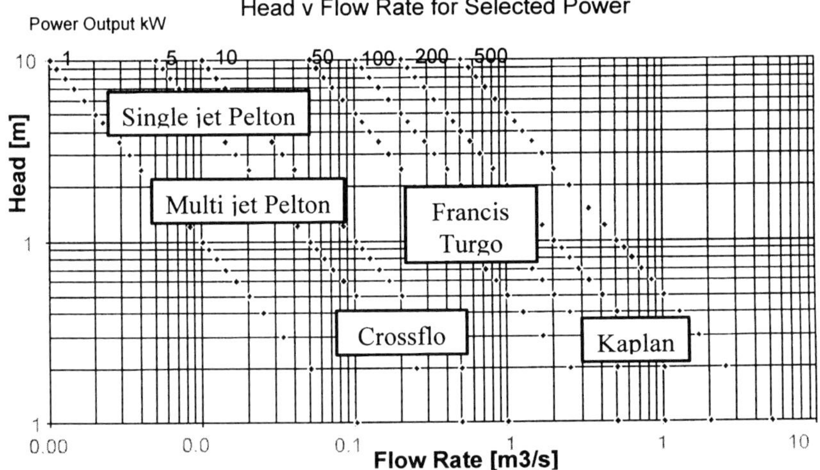

Figure 2. Impeller selection for Micro-hydro devices

For the students taking part in the projects the studies illustrate the issues in selecting between different design concepts – in alternative energy there is often a greater degree of choice of approach than in other engineering domains such as automotive engineering - and also introduce them to questions of patents and IPR.

3.3 Design Architecture and Selection of Parameters

Once a particular design concept has been chosen, then the embodiment phase of the design process involves choice of key parameters – sizes and arrangements – and configuration of the artefact. In the exercises that have been carried out this has involved issues of:

Choice of product architectures to allow greatest flexibility in adapting a basic device design to different circumstances, as for example in the selection of cross-flow turbine sizes that give the greatest flexibility in using a single micro-hydro design in different flow and head conditions. Figure 3 shows the choices made by Holden (7), in his choice of design parameters in this application.

Head (m)	Outer Diameter (m)	Rotational Speed (rpm)	Gear Ratio	Total Runner Length (m)	Total System Efficiency	Power Generated (W)	Section Lengths (m)	Number of Runner Sections	
1.5	0.10	490	3.1	1.5	50	1138	0.5	3	Type A
2	0.10	566	2.7	1.0	50	1168	0.5	2	Type A
2.5	0.10	632	2.4	0.5	50	816	0.5	1	Type A
3	0.10	693	2.2	0.5	50	1072	0.5	1	Type A
3.5	0.13	576	2.6	0.25	50	878	0.25	1	Type B
4	0.13	615	2.4	0.25	50	1073	0.25	1	Type B
4.5	0.13	653	2.3	0.25	50	1281	0.25	1	Type B
5	0.13	688	2.2	0.25	50	1500	0.25	1	Type B
5.5	0.14	670	2.2	0.125	50	932	0.125	1	Type C
6	0.14	700	2.1	0.125	50	1062	0.125	1	Type C
6.5	0.14	728	2.1	0.125	50	1197	0.125	1	Type C
7	0.14	756	2.0	0.125	50	1338	0.125	1	Type C
7.5	0.14	782	1.9	0.125	50	1484	0.125	1	Type C
8	0.14	808	1.9	0.125	50	1635	0.125	1	Type C

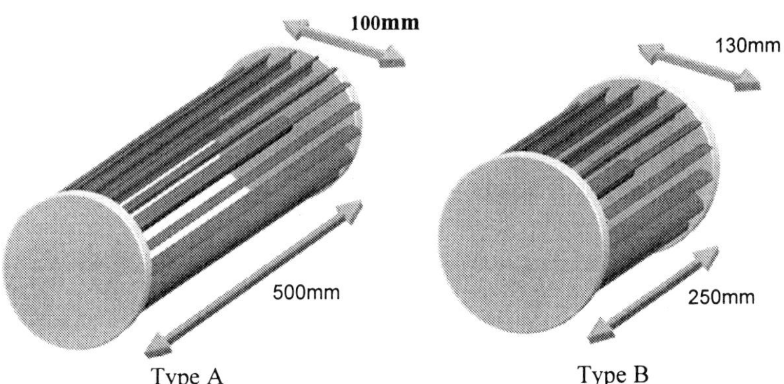

Figure 3. Modular architecture of cross-flow impellers for micro-hydro

- Development of mathematical and computational models of devices in order to carry out parametric sizing exercises. For example, the sizing of oscillating water column wave energy devices involves solution of the complex relationship between wave conditions, turbine size and the dimensions of the oscillating water column. Philips (8) developed a computational model of OWC devices in order to select the key parameters for devices to be used in wave energy farms off the south-west of England. Figure 4 shows an example of part of the output from this model.

3.4 Detail design

The detail design phase for the chosen devices introduces many interesting design issues, in particular concerning the selection of components such as bearings and seals for use in arduous conditions (e.g. for micro-hydro devices), the detail design of drive shafts for each of the devices, the design of device foundations, attachments and fixtures and so on. Issues of adaptation of devices to natural conditions – such as flow control for hydro plant or coping with differing wind-speeds for wind turbines – appear to offer considerable scope for interesting detail design projects in the future. Achieving good control at low cost appears to be a particularly important issue.

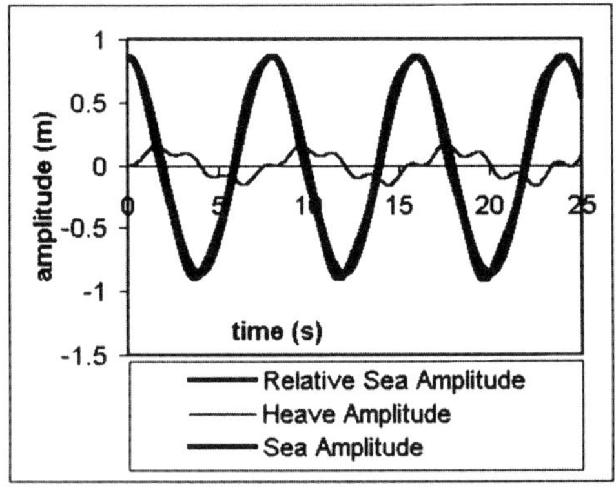

Figure 4. Model of OWC wave energy device heave performance

3.5 Product design issues

In addition to the engineering design issues involved with alternative energy devices, there are also the product design issues concerned with making them attractive to potential customers. This is particularly the case with small-scale devices designed to be sold to the general public – for example how can small-scale wind turbines be made attractive to domestic users? This question led to examination by the students studying wind energy of such issues as:

- The interface of the device to the domestic energy system, including selection of inverters and financial issues in selling energy to electricity suppliers.
- The use of the devices for specific purposes, for example for battery charging for electronic systems, for emergency power supplies, or for use in recreational activities such as camping.

3.6 Selection of Materials and Manufacturing Process

One of the central premises of the projects that have been carried out is that in order for alternative energy devices to become economically viable their price has to reduce compared with traditional energy sources. This may be achieved by exploiting mass production techniques and other approaches leading to economies of scale. Specific materials and manufacturing process issues studied have included:

- The selection of manufacturing process for OWC buoys for wave energy farms. This work has compared steel fabrication, casting and reinforced concrete construction for the main structure of the buoys.
- Identification of materials and manufacturing processes for the cowling, housing and blades of small wind turbines, including a comparison of pressed steel and various composite moulding techniques.
- Comparison of fabricated lattice steel, tubular and stayed tubular masts for small wind turbines, and identification of likely mass production costs for fixtures and fittings of such masts.

3.7 Design for Manufacture

Design for manufacture work has also involved a study of the application of design for manufacture considerations. The question that the students were asked to address was: what might be the benefit that might be obtained by manufacturing small alternative energy devices using the same manufacturing approaches applied to domestic power tools, white goods and automotive products. Work that is currently going on involves investigating the following DFM techniques applied to small wind turbines: part count reduction, use of standard parts, multi-functional parts, single direction assembly and so on.

4. OTHER EXAMPLES OF PROJECTS

Sustainable technology issues have also been good sources of projects in other design teaching areas. Other recent projects have included investigation of the energy requirements for manufacturing processes and design information collection on approaches to human-powered devices.

5. CONCLUDING REMARKS

Alternative energy devices have proved to be a fruitful source of design topics that allow a wide range of design issues to be explored. The students who have carried out the work have found the topic interested, and have been well motivated. It is suggested that the subject domain can also usefully be incorporated more widely into the engineering curriculum. For example, a fluids laboratory might examine the performance of different impeller arrangements for hydro-power, or a heat transfer laboratory might examine the radiation heat transfer in a solar collector. Engineering science subjects can also make wide use of sustainable technology issues as a source of examples – e.g. velocity vector diagrams for water turbines rather than pumps, efficiency of wind turbines rather than propellers, thermodynamic cycles for wind-driven heat pumps rather than refrigerators or manufacturing energy costs rather than financial costs. If sustainable development is as important as many believe then universities and colleges should lead the way in promoting awareness, not just in dedicated classes, but across the curriculum.

ACKNOWLEDGEMENTS

The author wishes to thank in particular the students that have worked with him over the years on the projects described, including Eva Assayag, Helen Barker, Chris Holden, Joe Phillips, Iain Smith and Dave Williams. Their enthusiasm for the subject has made working in the area a pleasure.

REFERENCES

(1) Graedel, T.E. and Allenby, B.R., "*Industrial Ecology*", Prentice Hall, 2002

(2) Bryant, E., "*Climate Process and Change*", Cambridge University Press, 1997

(3) Gipe, P., "*Wind Energy Basics: A Guide to Small and Micro Wind Systems*", Chelsea Green Publishing Co., White River Junction, Vermont, 1999

(4) Falnes, J., "*Ocean Waves and Oscillating Systems: Linear Interactions Including Wave Energy Extraction*", Cambridge University Press, Cambridge, 2002

(5) Harvey, A. and Brown, A., "*Micro-hydro Design Manual: A Guide to Small-scale Water Power Schemes*", ITDG Publishing, 1993

(6) Goswami, Y. D., Kreith, F., Kreider, J. F. and Brown, R. C., "*Principles of Solar Engineering*", Taylor and Francis, 2000

(7) Holden, C., "*Small-scale Hydro Power Devices*", University of Bristol, Department of mechanical Engineering, 2001

(8) Philips, J.L., "*Simulation model of OWC wave energy devices*", University of Bath, Department of Mechanical Engineering 2003

Design for self powered, self disassembly of vehicles

N JONES, D HARRISON, H HUSSEIN, and **E H BILLETT**
Department of Design, Brunel University, Egham, UK

ABSTRACT- ACTIVE DISASSEMBLY

It was found that in the context of automobile recycling, there is enough residual energy in end of life vehicle batteries to trigger many self-disassembly devices. An analysis was completed to investigate the residual energy levels remaining in end of life vehicle batteries. A design table was then constructed to allow designers to 'design -in' a variety of devices that will reliably trigger without exceeding the overall energy level contained within the battery. At least 143 Shape Memory Alloy devices made from 25-micron diameter wire, or 41 50-micron devices can be triggered, or combinations thereof. This provides the possibility of designing and creating self-powered, self-disassembling products for recycling.

1. INTRODUCTION

Active Disassembly Using Smart Materials (**ADSM**) is a concept that has been developed at Brunel University. It is a system allowing assemblies to readily separate for recycling, when they are exposed to certain triggering conditions. Shape Memory Alloy (SMA) actuators can be incorporated into product housings. When the product is heated above the SMA's transition temperature, the SMA actuator undergoes a shape change *(the 'Shape Memory Effect or SME)*, that exerts a force. This force is used to break apart a product casing. Now with the implication of the 'End of Life Vehicle Directive', we have expanded our research to investigate automotive disassembly.

Systems of retainers were constructed to eject the battery of a mobile telephone when the end of life condition arises. There were two novel elements in this work.

Firstly, this was the first application of electrically triggered ADSM. By utilising SMA 'muscle wires', actuation systems were developed that would trigger not by external heat application, but by passing a current through the wire. It is the current flow that provides the necessary rise in temperature for the SME to occur. In this instance the wire contracts with enough force to eject the battery from the host assembly. Muscles wires have been widely used in many applications before, including heat engines and novelty items (1), but this is the first example of their use for *non-destructive* product disassembly.

Secondly, this was also the first example of ADSM occurring with no energy input from outside the product. Previously, thermal energy has been applied to the assemblies by various means including microwave heating, induction heating, hot air convection, hot water immersion and direct heating with a hot probe. In this instance, it is only the electrical energy contained within the assembly that is utilised.

Smart materials have been investigated in the Department of Design at Brunel since the mid nineties. The investigations centred on design for sustainability and design for recycling aspects (2). The most successful area for investigation has been ADSM, which is now the subject of an EU Framework V investigation project between the University and some major industrial partners from the consumer electronics industry. Given the success in this area, independent research continues to investigate how this technology could be incorporated into the automotive sector. With the implementation of the EU ELV directive (3), automotive recycling has never been more topical. This paper describes how an initial experiment into mobile telephone recycling has progressed to explore the concept of self-powered self-disassembly of automobiles.

2. SMART MATERIALS OVERVIEW

For Active Disassembly purposes, the materials that have been used on previous projects can broadly be classified into two categories, Shape Memory Alloys (SMA), and Shape Memory Polymers (SMP). These materials change their mechanical properties in relation to temperature, or more specifically, a 'transformation temperature', 'T_x'. When T_x is reached, the 'Shape Memory Effect' ('SME') occurs and the material will physically deform or recover to a previously formed shape. For this investigation, only Shape Memory Alloy Solutions have been used.

2.1 Shape Memory Alloys

The SME is a unique property of certain alloys exhibiting martensitic transformations. Even when a shape memory alloy is deformed in its low temperature phase, it can recover its original shape by reverse transformation when heated to its critical transformation temperature. The same alloys have another unique property called superelasticity (SE) at a higher temperature, which is associated with a large (up to 18%) non-linear recoverable strain upon loading and unloading. Since these alloys have a unique property in remembering their original shape, they can be used in various functions such as actuators, couplings and medical implants (4).

Shape memory alloys are capable of exerting significant amounts of force when they are stimulated to change shape. Using this force, a piece of SMA material is therefore capable of being formed into an actuator. This actuator can be triggered and used to disassemble a host product or assembly. For example, in the case of the disassembled calculator, coiled CuZnAl actuators were concealed within the product. When the actuators underwent their SME, the products snap-fits were broken causing disassembly. Upon cooling the actuators returned to their previous state (Figure. 1) (5). Similarly, SMA devices can exert significant amounts of restraining force when they contract and undergo their SME. SMA devices will remain stable and inert in a host product until the required triggering conditions occur.

Figure 1. A Calculator Actively Disassembled Using Concealed SMA Actuators

3. INITIAL ELECTRICAL INVESTIGATION USING A MOBILE TELEPHONE

A further refinement of the Active Disassembly concept has recently been investigated. Systems of retainers were constructed to eject the battery of a mobile telephone when the end of life condition arises (6). There were two novel elements about this work.

The power able to be delivered by the batteries after discharging them was between 1.1 and 1.5 watts. From the wire activation experiments we have seen that this is enough to trigger at least 4 similar devices at EoL. Alternatively, there is enough energy to reliably provide at least 45 seconds of continual actuation.

To more usefully use this energy, it would certainly be more beneficial to trigger multiple devices. There is also the possibility of introducing hierarchical disassembly of the telephone to allow piece by piece dismantling of the product (7).

3.1 Further Work, and Levels of Innovation.

These results show the potential for using residual charge in a battery-powered product to provide energy for self-disassembly. In a design for recycling context, self-disassembly of bulk quantities of products would provide a valuable reduction in manpower and remove a bottleneck in the recycling process. The technology would be utilised most efficiently in conjunction with the disassembly of the complete telephone, in a 'disassembly for recycling' context.

There is the potential with this technology of taking further innovative design steps. With the actuator being electrically triggered, there is the possibility of activating actuator devices within the product through a keypad. Technical staff at repair/reprocessing centres may use the keypad interface to trigger 'active' devices to enable disassembly or removal of any subassemblies for servicing, repair, or refurbishment.

In a hierarchical disassembly scenario, the disassembly would be sequential at specific points within the recycling stream. During hierarchical disassembly, separating the telephone fascia, the screen, then the printed wiring board and the battery will produce four separate waste streams for reprocessing. This method also provides a simple solution for disassembly of the

telephones where the battery is effectively 'trapped' inside the telephone, for example, the Nokia 8210.
The falling cost of Nickel-Titanium in recent years makes using these devices more feasible than in the past. However there is a need to assess the economic consequence of the incorporation of this technology as opposed to other more traditional disassembly methods.

4. AUTOMOTIVE APPLICATIONS

Following on from these experiments, a second energy assessment was completed to evaluate the residual energy in End of Life Vehicle (ELV) batteries. Similar to the telephone scenario, when a car battery will no longer turn a starter motor, or power 120 watts of Halogen headlamp, it does not necessarily mean that there is no longer any useful energy left in the battery. Indeed, given the size and capacity of vehicle batteries, the energy in 'dead' car batteries could be orders of magnitudes larger than the energy contained in hand held electronic devices. We completed an evaluation of the energy contained within an EoL vehicle battery to see if it could be usefully harnessed.

4.1 Experimental Method

As we have already proved the potential for electrical energy to be used to trigger ADSM devices, an overall energy evaluation of 30 ELV batteries was conducted.

Thirty ELV batteries were obtained from an automotive scrap-yard. There was no specific selection process for the batteries. The batteries were collected; some in a visually good state of repair and some that were showing bad signs of damage and leaking. As these batteries represented a typical cross section of vehicles at end of life, all of the batteries were used and none discarded. The batteries had been standing in the scrap-yard for an unknown period of time, and were tested over the course of one month.

A 9V PP3 (*Duracell MX 1604*) battery was discharged through a 150-Ohm resistor (actually measured at 148.7-Ohms) to act as a control experiment. This gave an acceptable power dissipation of 0.624 Watts with the initial starting voltage of 9.63 Volts.

Each vehicle battery was connected in series and discharged through the same 20Watt 148.7-Ohm resistor. The Voltage against time was plotted using a chart recorder, and a power curve was calculated from this. Integrating the area under the power curve represented the energy contained with the battery.

4.2 Results

The energy contained within the ELV batteries was clearly significant compared to the telephone batteries.

The 9v PP3 battery provided significant energy for nearly 13 hours of discharge, at which point the Voltage suddenly dropped from ~4.5V to ~0.5V. The battery was no longer able to deliver any current after 18hours. This was consistent with the Duracell technical data (8).

When compared to a large automotive load (i.e. headlamp etc), the energy required to drive a shape memory device is comparatively low. This therefore meant that the batteries were discharging for several weeks before any changes in the batteries ability to deliver any current were noticed. It was therefore decided to simply test the behaviour of each battery over a 24-

hour period. Voltage and current were measured while the battery was discharged through the 150-Ohm resistive load. These measurements were then integrated to calculate the total residual energy.

The range of energy contained within the EoL batteries was between approximately 1/2 kJ up to 60 kJ. The results show that even in the worst case scenario, a severely discharged and leaking battery with a terminal voltage starting at 1.79 volts, falling to 0.17 volts, contained 588 Joules of useful energy.

4.3 Conclusions

The discharge curve of the batteries was as expected, in that the discharge rate remained very linear (9). The control battery also discharged consistently with the published data sheets. From the results we could see that for all of the 30 batteries there is more than enough energy to trigger shape memory devices similar to those used in the mobile telephone. Given the extra energy available in the waste automotive batteries, we can see from table 1 that many more active devices could be used.
The least healthy of the batteries tested contained 588 Joules of energy, and it is based on this battery that the information overleaf was calculated.

Table 1. An Illustration of How Many Shape Memory Devices of Each Size May Be Reliably Triggered by an EoL Vehicle Battery

SMA Wire Diameter/ microns	Resistance / Ohms m^{-1}	Current Drawn at 12V (100mm wire length)/ Amps	Energy consumption/ Joules sec^{-1}	Energy needed for 5 seconds of triggering/ J	Minimum number of devices able to be triggered at EoL. *(Whole No's.)*
25	1770	0.068	0.818	4.09	143.77 (143)
37	860	0.140	1.686	8.43	69.75 (69)
50	510	0.235	2.816	14.08	41.76 (41)
100	150	0.800	9.600	48.00	12.25 (12)
150	50	2.400	28.800	144.00	4.01 (4)
250	20	6.000	72.000	360.00	1.63 (1)
300	13	9.230	110.769	553.85	1.06 (1)
375	8	15.000	180.000	900	0.65 (0)

It must be pointed out that the worst battery in the assessment (battery no. 5) provides the information above in theory only. The battery was in such a poor condition that it was unable to deliver the current required for the larger (250 micron and up) wires. However, this battery aside, the rest of the batteries were able to trigger any of the other wires.

The information presented in Table 1 provides a very useful design tool to aid the incorporation of ADSM technology into battery powered devices.

It was also noticed from the results, that for a few batteries the potential difference measured at the terminals actually increased during the test cycle. It is well known that the working voltage of a practical cell under load can rise with an increase in temperature. This is due to the drop in internal resistance of the cell caused by increased conductivity of the electrolyte phase (10). However, given the amount of energy contained within the batteries, this small rise is comparatively negligible.

4.4 Further Electrical Development.
Electrically activated active disassembly devices are not limited to muscle wires. Other shape memory fasteners and actuators that have been previously thermally activated could be triggered electrically if a small heating element or coil is built into the fastener (Figure 2).

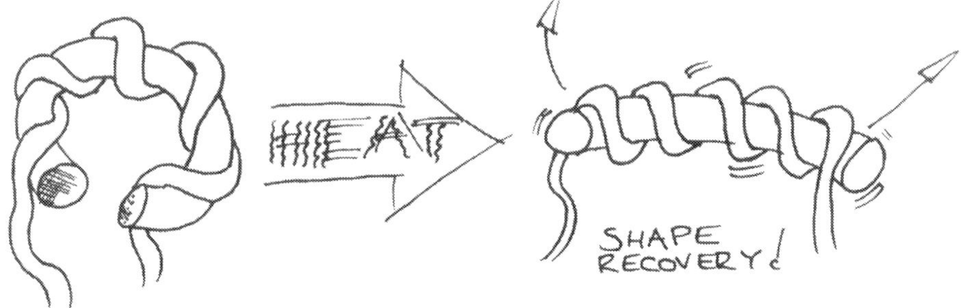

Figure 2. Electrically Heated SMA Clip

Utilising again the residual energy in the waste automotive batteries, we produced prototype clips triggered by electrically heated Nichrome wire. The wire must be coated to prevent the current in the clips earthing through the actual shape memory devices. A PTFE coated wire was found to be the most suitable as it was able to withstand higher temperatures (up to 240^0c) than conventionally coated wires (Figure 3).

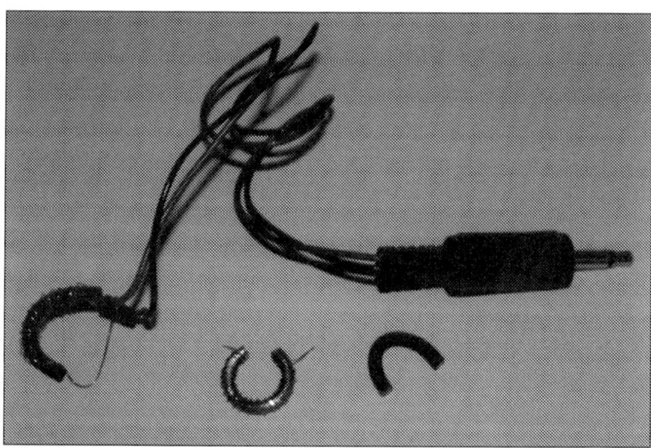

Figure 3. Electrically Heated and Unwound Circlip, Next to an Un-triggered Clip

4.5 ELV Applications
In an automotive context, this inspires the possibility of using the residual charge in the car battery of an ELV for powering the disassembly of the host vehicle. In this scenario, active disassembly can be achieved when the vehicle is being depolluted. In conjunction with the ELV legislation, all vehicles will be depolluted at their end of life. In this scenario, there is a window of around 20 minutes when the vehicle is on the ramps at the depollution station

being drained of fluids. It is at this point that the shape memory devices could be triggered. Once the devices have been triggered and the appropriate parts have been removed from the car, the battery can then be finally removed from the car and recycled or reused.

Current recycling figures put automotive recycling at a figure of around 75% *(by mass)*. This is still higher than any other industrial sector. However, this is still 10% short of the 2006 target. Active disassembly technology could allow easy and economic recovery of additional automotive parts, making directive compliance easier.

REFERENCES

(1) Perkins, J., and Hodgson, D., '*The Two Way Shape Memory Effect*', In, '*Engineering Aspects of Shape Memory Alloys*', Ed Deurig, T. W., et al. Butterworth- Heinemann 1990, Pp 195-206

(2) Chiodo, J. D. et al. '*Active Disassembly Using Smart Materials (ADSM) Introductory Materials Overview.*' In, Proceedings of the Fourth International Conference on Ecomaterials, Gifu, Japan. The Society of Non- Traditional Technology, 1999

(3) Commission for the European Communities. Proposal for a Directive of the European Parliament and the Council on End of Life Vehicles 1997/0194 (COD)

(4) Otsuka, K., and Wayman, C. M., '*Shape Memory Materials*'. Cambridge University Press, 1998 Pp2-3

(5) Chiodo, J. D., Jones, N., Harrison, D. J., Billett, E. H., '*Shape memory alloy actuators for active disassembly using 'smart' materials of consumer electronic products*'. In, Journal of Materials and Design, Vol. 23 Issue 5. Elsevier Science 2002.

(6) Hussein, H., Jones, N., *et al*, '*Design for Active Battery Removal in Mobile Telephones*'. In, Proceedings of the International Congress on Battery Recycling, ICBR 2002- Vienna, Austria, 2-5 July 2002. ICM 2002.

(7) Chiodo., et al, '*Investigations of Generic Self Disassembly Using Shape Memory Alloys*', *In,*IEEE 1998: International Symposium on Electronics and the Environment, 4-6 May, Oak Brook, Illinois. Pp. 82-87.

(8) Duracell technical data (Alkaline cells)- http://www.duracell.com/oem/Primary/Alkaline

(9) Tuck, C. D. S. (ed.) '*Modern Battery Technology*'. Ellis Horwood 1991. Pp.204-207.

(10) Vinvent, C. A., *et al*, '*Modern Batteries*'. Edward Arnold 1986, ISBN 0713134690. Pp-30

Using the recycling theme to motivate product design students – a teaching methodology based on domestic can crushers

S PACE
Department of Design and Technology, Loughborough University, UK

ABSTRACT

A can crusher design exercise has been used to investigate the design strategies adopted by first year undergraduate students with different academic backgrounds. The techniques of reflective design analysis have been used to examine the work of two different groups, one group having a science-based background and one group having an arts-based background. The analysis has identified a polarisation of strategies within the two groups that reflects the academic background. The research findings have been used to develop a structured teaching methodology that overcomes the perceived limitations of the school background and provides both types of student with broad-based problem solving strategies.

1. INTRODUCTION

Designers are increasingly being encouraged to examine the balance between morality and economics in their work. A requirement for environmental accountability has become a feature of both the moral and legislative framework within which designers of the twenty first century are constrained. The need to conserve natural resources now features prominently in the economic strategies of the industrialised nations and the recycling of metal products is a fundamental pillar of these strategies.

No metal product is more ubiquitous than the metal can and the economic case for recycling cans is overwhelming. However, a large proportion of metal cans continues to be dumped as a consequence of indifference in the domestic household and the costs involved in transporting large volumes of waste. Both of these problems are long term issues that may ultimately require to be addressed by legislation. In the short term, it is possible to reduce the problems by providing households with convenient means to crush and store cans.

Can crushing products are already available, but many prospective purchasers are not persuaded to buy them. Sales could be improved if the products could be made more affordable, desirable and interesting without compromising their functional effectiveness. These requirements provide a stimulating brief for young designers who are generally more environmentally aware than their forebears by virtue of the prominence given to such issues in the school curriculum (1). Young designers can be motivated by the desire to shape the future and increasing numbers are pursuing career ambitions in design by enrolling on combined product design and engineering degree programmes (2).

The combined programmes offered at prominent universities recruit high calibre school leavers, but applicants are recruited with different academic subject backgrounds. Some applicants are able to offer 'A' level subjects that qualify them for traditional engineering degree programmes, whilst other applicants gain entry by showing design aptitude in art-based subjects. Research indicates that individuals within these two groups bring with them particular modes of cognising and learning styles that reflect their academic subject choices in the final years of school (3). This poses problems for undergraduate course organisers in providing design-based learning experiences that can accommodate and build upon the different learning styles. If course organisers are to provide a broad, practically-based education for product designers (4) who are able to combine the skills of the industrial designer and the design engineer, then the pedagogic issues with regard to learning styles must be addressed from the outset.

A can crusher design exercise has been devised and used to investigate the strategies adopted by first year students from both types of background. The techniques of reflective design analysis have been used to examine students' work and the analysis has identified a polarisation of strategies within each group. Problem-focused strategies have been found to be prominent at one extreme and product-focused strategies prominent at the other. The research findings have been used to develop a structured design methodology that integrates these disparate strategies and identifies the teaching elements required to support the structure.

2. A STRUCTURE FOR DESIGN PROJECTS

Previously acquired learning styles can limit an individual's ability to learn the design strategies that are needed to solve product design problems (5). Where technical product performance is important, student designers need opportunities to develop learning styles that enable them to manipulate both engineering and aesthetically based concepts (6). A suitably structured methodology for identifying such opportunities within an integrated approach to the design of products for technical performance is given in Figure 1.

Figure 1 shows how the product design process can be treated as a logical progression from problem definition to problem solution and identifies the individual elements of the process so that the implications for learning can be drawn out. The individual elements can be concerned with mathematically-based concepts or art-based concepts and broad-based product designers need to develop learning styles that are appropriate in both areas. In order to promote development of appropriate learning styles, exercises can be set which place emphasis on specific elements. Exercises set within a well-bounded context provide purpose and relevance to learning. Where exercises are sufficiently broad-based all the necessary design perspectives and the learning issues within individual elements can be addressed.

Figure 1 also provides a structure for the generation of ideas to solve problems in technical product innovation. The structure consists of four parallel strands of design activity at five different levels and shows that ideas can spring from any individual area. The individual areas of design activity are shown as elements covering the necessary spectrum for generating new ideas for mechanically functional consumer products. The structure shows how new ideas can be generated from any one of four different perspectives - mechanisms, styling, ergonomics and manufacturing. In theory, an infinite number of feature permutations can be made which suggests that an infinite number of design ideas can be generated to satisfy a single design concept.

However, the number of ideas that can be generated depends on the extent of the knowledge base in each of the four areas. The development of an extensive knowledge base depends in turn on the ability of the student to apply the learning styles characteristic of each area. At one extreme the student is required to learn analytical techniques concerned with understanding and developing mathematical and technical models and at the other extreme the student is required to learn drawing and modelling techniques concerned with developing product aesthetics.

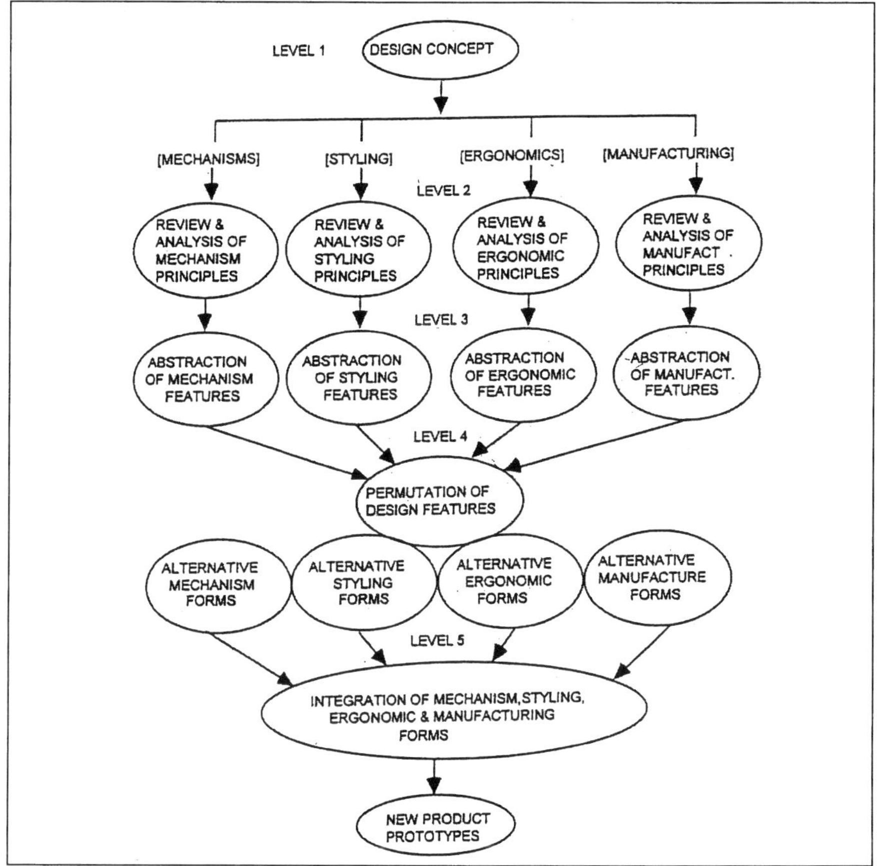

Figure 1. A structure for design projects

3. USING THE STRUCTURE FOR PRODUCT ANALYSIS

A better understanding of any design problem can be derived through a practical exploration and analysis of previous attempts to solve the problem. Where these attempts are in the form of existing products, the structure given in Figure 1 can be used in reverse to identify the design decisions made in formulating the particular mechanical arrangements and product features of different designs. For example, an analysis of the performance characteristics of the proprietary can crusher shown in Figure 2 indicates that the device has a suitable characteristic for axially crushing cans – the mechanism produces a toggle action at the start and finish of the stroke where the maximum forces are most required.

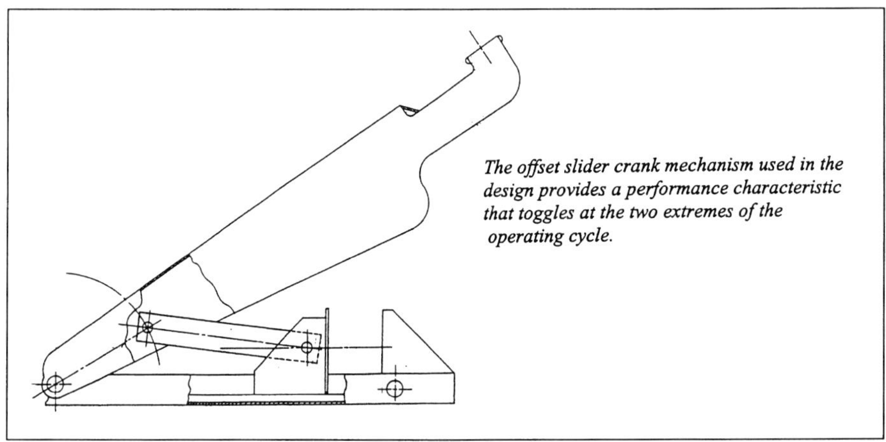

Figure 2. Mechanical arrangements in a proprietary can crusher

The machine has been designed to crush tall drinks cans, but is not nearly as effective for crushing shorter heavier gauge pet food cans. The task requirements for the shorter cans is shown in Figure 3 which indicates that peak forces of some 3kN are required for initiating crushing and final compacting of a pet food can.

In analysing just one device, the process of bringing together analytical models of task and technical performance requires the practising of particular styles of learning and modes of cognising. Once learned, these styles of learning can be applied in investigating the potential of other mechanisms. Many other proprietary products achieve high force magnifications using the toggle principle in various configurations and with a little imagination some of these configurations may also provide ideas for solutions to the problem of designing a can crusher.

Similarly, product analysis techniques (7) applied in the remaining design perspectives can provide practice in applying the cognitive and practical modelling skills that need to be integrated into the process of designing products. The styling and ergonomic features are important attributes of products that play a significant part in consumer choice and these features are in turn influenced by manufacturing considerations. The methodology allows a focus to be made on learning objectives in any particular perspective and cultivates appropriate learning styles across the boundaries of different design perspectives.

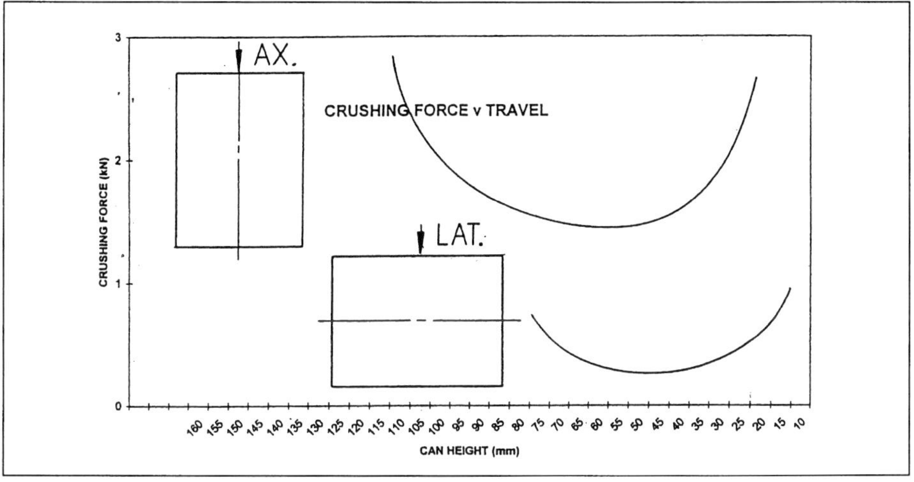

Figure 3. Task characteristics for axial and lateral crushing of a pet food can

4. USING THE STRUCTURE TO EXAMINE THE SCHOOL BACKGROUND

In order to examine the influence of the school background on the learning styles and design strategies used by students in tackling technical design problems, a can crusher design project was set without the students being given a structured methodology. The work of two different student groups was examined. The first group consisted of students with a strong 'A' level background in technical subjects with career ambitions in mechanical engineering and the second group consisted of students with a strong background in art-based subjects and career ambitions in product design.

The students devised and conducted their own testing and product analysis programmes and used the experience gained to generate design proposals. The structure shown in Figure 1 was used to analyse the strategies employed by both groups and Figure 4 was produced to illustrate the process of analysing the efforts of three product design students. The three examples shown are typical of product design students and provide examples of designs where the screw thread mechanism was chosen as the operating principle.

The examples show a product focused approach where the need to produce a 'product' has been the main influence on the strategies adopted. None of the designs are suitable for generating the large forces necessary for crushing pet food cans - screw thread theory could have shown the limited potential of such designs. The product features have been based on the theme of pets and all the designs provide some visual interest. The manufacture of the products has been considered and the details are clearly derived from an analysis of the three engineering product assemblies shown.

The strategies adopted show the influence of the school background in that each student has applied previously learned strategies that limit their efforts to just three of the four design perspectives. Each student has essentially ignored the most demanding technical

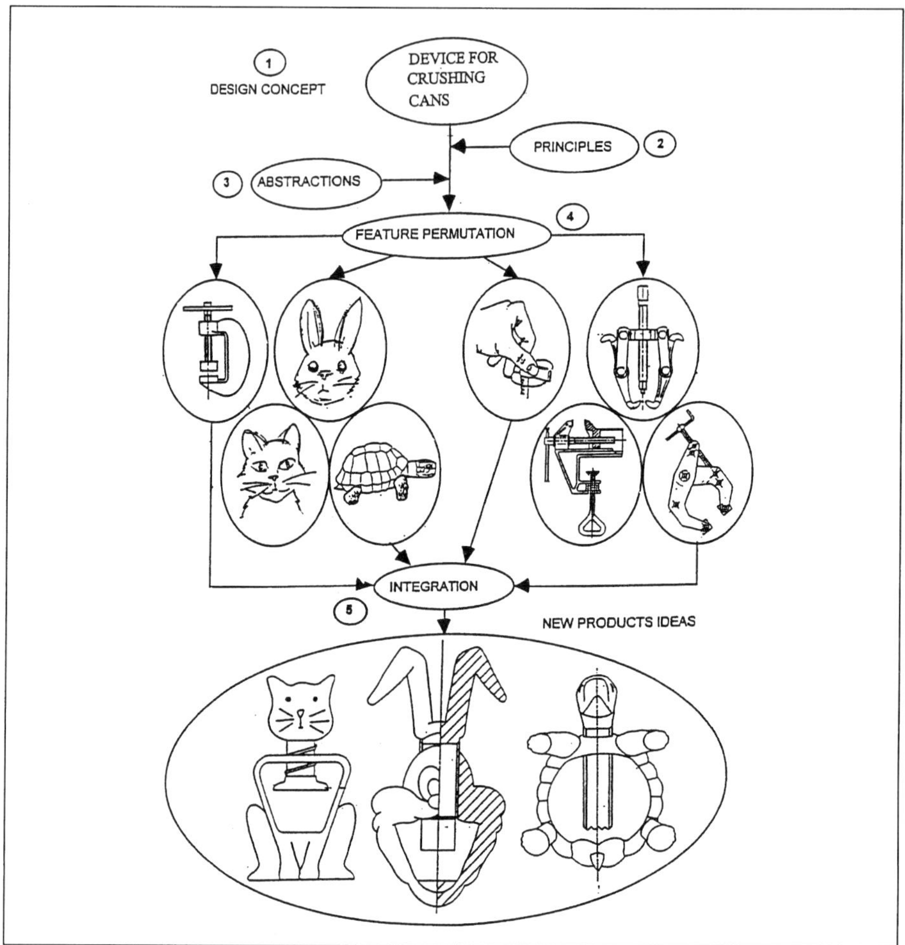

Figure 4. Using the structure to examine student design strategies

performance aspects of the problem and has proceeded to tackle a lesser problem ignoring the need for quantification. The limitations of the learning styles derived in school have precluded any attempt by the student to learn by themselves the necessary elements of mechanical science to analyse both task and mechanical devices. However, the designs illustrate that a base of skills and practical experience has been applied in the generation of ideas and that the skills base consists largely of sketching, drawing and modelling techniques.

Similarly, the engineering students have applied skills and experience in generating solutions and Figure 5 provides an example of a student engineer's design which is also based on the principle of the screw thread. The device is clearly intended to satisfy the requirement to generate a high force magnification and the student has concentrated his efforts only on producing a solution to this aspect of the problem. This solution is derived from a problem focused approach and is typical of the solutions offered by engineering students.

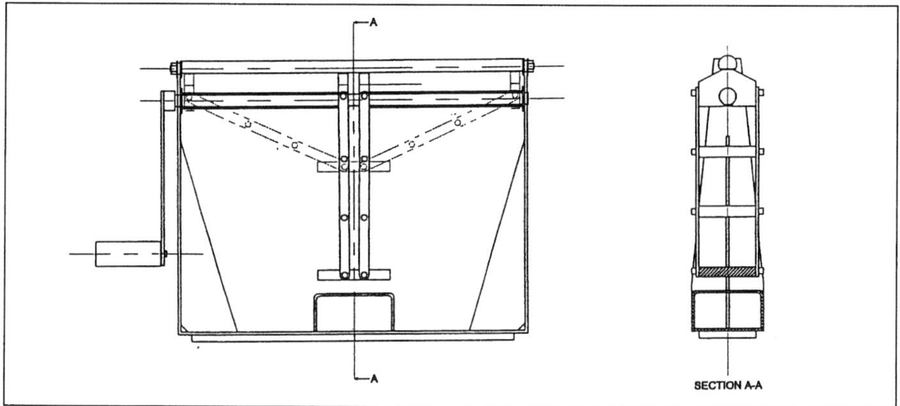

Figure 5. A problem focused design

The solution reflects the school mathematics background where solutions tend to consist of single answers, rather than the generation of mathematical models (8). The device satisfies a single requirement and the linkage incorporated in the design is inappropriate for satisfying the task characteristic of high force magnification at both start and finish of the crushing action.

The device shown is not likely to form the basis of a successful product, although it demonstrates a limited form of success in mechanical design. The student's previously acquired learning styles have prevented the application of a design strategy based on a properly quantified analysis of task and mechanism characteristics. The student has attempted to satisfy requirements in three out of the four necessary design perspectives, neglecting any requirements to produce a product that is also desirable and interesting. The example shown is typical of the efforts of student engineers and shows that the school background lacks the sketching and drawing skills necessary to explore ideas in product aesthetics.

5. CONCLUSIONS

Designing inexpensive consumer products capable of delivering large forces within a prescribed cycle of operation and making such products desirable and interesting is problematic. Figure 3 shows how the requirements for crushing strong steel cans may be made more manageable by using a lateral crushing action on cans which have had both ends removed. Consumers could not reasonably be expected to remove both ends of a typical pet food can, but if manufacturers could be persuaded to provide ring pulls, or some such device, at both ends of the can, then the task of crushing becomes much less demanding. Suitably designed cans would provide greater opportunities for product designers to produce interesting can crushers that people would be prepared to buy. For example, large food and drink manufacturers invest heavily in brand identities, product images and company logos which could form the styling inspiration for young people's design efforts. Also, some charities rely on the income from can collections and organisations including the RNIB, RSPCA and RNLI have their own public images that could be used to inspire young designers.

The work has shown that recycling issues can be used to motivate young people and provide a worthwhile outlet for creative effort. Recycling issues can be used to generate broad-based design tasks – tasks which require designers to exercise self-sustaining learning styles and to work in different design perspectives. These perspectives cross traditional academic boundaries and the methodology presented in the work enables teachers to identify subject specific learning issues and to integrate teaching within these specific areas into a coherent structure. The structure gives direction and focus to problem solving efforts in the field of consumer product design and provides a simple design strategy that can be learned and applied by students.

Both groups of students examined in the study were found to employ similarly deficient strategies and it is concluded that these deficiencies were a consequence of the learning styles developed in school through the study of particular academic subjects in the final years. It was also found that these previously acquired learning styles inhibited creativity by limiting the range of design perspectives that could be addressed in the search for solutions.

The methodology presented in the work has been devised to overcome the perceived limitations of the school background. The methodology provides novice designers with design strategies that can be applied across the different perspectives of broad-based consumer product design. The methodology has been used to address an issue in materials recycling, but is generally applicable where the objective is to teach students how to design mechanically functional artefacts that are also desirable and interesting.

REFERENCES

(1). Department for Education, *"Design and Technology in the National Curriculum"*, HMSO, London, 1995.
(2). Universities and Colleges Admissions Service (UCAS), *"Annual Reports 1996 – 2000"*, Cheltenham.
(3). Woolnough, B. E., *"The Making of Engineers and Scientists"*, Oxford University Department of Educational Studies, Oxford, 1991.
(4). Ulrich, K.T. and Eppinger, S.D., *"Product Design and Development"*, McGraw Hill, New York, 1995.
(5). Lawson, B.R., *"How Designers Think"*, Architectural Press, London, 1980.
(6). Muranka, T. and Rootes, N., *"Doing a Dyson"*, Dyson Appliances Limited, Malmesbury, England, 1996.
(7). Baxter, M., *"Product Design"*, Chapman & Hall, London, 1996.
(8). Pace, S., *"Mathematical Modelling for Students of Product Design: an Example from the National Curriculum in the United Kingdom"*, Teaching Mathematics and its Applications, Vol 16.No.3,1997.

Surrogate wheelchair users in user-centered design – an ethical question?

W F GAUGHRAN
Department of Manufacturing and Operations Engineering, University of Limerick, Ireland
E H BILLETT
Department of Design, Brunel University, UK

ABSTRACT
The use of surrogates in user-based design, for wheelchair users (WUs) may raise some ethical questions. Goldsmith (2000) suggests that it is quite appropriate to use surrogate wheelchair users in establishing anthropometric and ergonomic data. Is the data gathered by such means valid? Is it ethical to use able-bodied people sitting in wheelchairs and say that the data gathered is applicable in designing for wheelchair users? At the University of Limerick, a number of WUs have been evaluated, and a like cohort of surrogate wheelchair users (SWUs), in establishing design ergonomics related to workbench design, with particular reference to best-fit working heights. Both groups were tested and comparatively analysed to ascertain factors-of-difference and to test the legitimacy of using such subjects in future design research. Optimum work heights, relating to comfort and efficiency have been established and the feasibility of using SWUs in determining best-fit ergonomics is discussed.
Key words/phrases: *Wheelchair/ Surrogate Wheelchair user ergonomics, Optimum working-heights, Inclusivity.*

1. INTRODUCTION
Ethics relating to design can have many facets, amongst which are, designer/client trust, materials selection, ecological and safety issues. However, there are issues of an ethical nature which impact at a more fundamental level. The establishing of data relating to design ergonomics may be one such area. As part of a research project at the University of Limerick, it has been necessary to develop ergonomic data relating to best-fit workbenches. As the project was for universal application, it was deemed necessary to include wheelchair users (WUs). WUs with normal upper body strength would be the most likely to use industrial workbenches, and such subjects can be difficult to locate. Goldsmith[1] says that it is appropriate to use surrogates in establishing ergonomic data, but there is no evidence that this is appropriate for industrial bench tasks. This element of the research project therefore makes a comparative analysis of a cohort of WUs and a like cohort of surrogate wheelchair users (SWUs), and is primarily concerned with best-fit bench ergonomics, with particular reference to optimum working heights.

1.1 The Test Cohorts
While the research concerns itself with trial fittings in a 'design-for-all' context, this part of the study sets out to determine best-fit work-heights for wheelchair users, at the workbench. The study was undertaken in three parts, (i) a cohort of paraplegic wheelchair users with

'normal' upper body strength. The criterion for suitability was that they should be capable of propelling themselves in the wheelchair, i.e. without the use of a motorised model; (ii) the same test bank for a similar cohort of surrogate wheelchair users; (iii) a comparative analysis of the findings for both cohorts.

2. EXPERIMENTAL PROCEDURE

2.1 The Test-bench Rig

In order to accommodate the individual requirements of the user and to determine a preferred height, efficiency rates, and estimated endurance, it was necessary to design a rig, which would cater for a range of heights. Pilot anthropometric data collected, determined that the lowest knee clearance required would be 570mm and a range of heights up to 200mm above Knee Height (KH) would be required. An electronically controlled, rack and pinion operated, telescopic lifting column was used to support the bench-top. To accommodate the depth requirement for knee clearance, the bench-top was cantilevered, and as the minimum column height was 610mm, a cranked support bracket was designed to allow for the lower levels. The work surface dimensions were deemed to be sufficient at 1200mm x 650mm. The finished test-rig had a height range of 570mm to 1100mm.

2.2 The Test Bank

A range of standard manipulative tests was chosen as well as a typical electric plug-top assembly. The tests and their associated values were as follows: Three Pin Plug Assembly (Total time to assemble three plug-tops); One Hole Test (The largest throughput in three trials in one minute); Small Tapping Task (Number of pairs of taps within the targets after 30 seconds); Purdue Pegboard (Total number of pins for left hand, right hand, both hands, and assembly); Grooved Pegboard (Total time to complete Pegboard); MRMT (Time to complete four trials); Large Tapping Task (Number of pairs of taps on the targets after 30 seconds).

Observation and WU interviews revealed that the most restrictive factor at traditional workstations was an inability to accommodate the user's legs underneath. This meant that the user could not work close to the bench in a comfortable working posture. The second element, which created significant difficulty, was the height of the working surface. Normally, the work-height for ambulant users is associated with elbow height. However as the usual difficulty for WUs was associated with knee clearance, it was decided that their knee-height would be used as the datum level. For the purpose of analysis, several anthropometric measurements were taken. These were: sitting stature, knee height, shoulder height and elbow height. The test-rig was relative to the subject's knee height but analysis of elbow height would were also made. The test-rig represented the traditional bench and five other heights, ranging from knee height to knee height plus 200mm, see Table 1 for heights range.

Table 1 – Test Heights

Height Number	Height Level
H 1	Knee Height (KH) - inaccessible
H2	KH + 50
H3	KH + 100
H4	KH + 150
H5	KH + 200
H6 (800mm)	Existing – inaccessible bench

2.3 The Test Subjects

Twelve wheelchair users were identified for the tests, but one was later eliminated because of insufficient hand-dexterity. The age range was from 19 years to 55 years. The period of time the test cohort were using wheelchairs ranged between 1.5 years and 37 years. Their disabilities resulted from stroke, spina bifida, and accident. The gender mix was seven males and four females. A similar test group of SWUs were selected.

2.4 The Test Method

2.4.1 Wheelchair Users Tests

All the subjects were briefed on the test objectives and completed a questionnaire. The duration of the test bank was approximately three hours. The Latin Square Order was used to randomise the test elements and the sequence of the bench heights. Each WU test-subject used their personal wheelchair. The anthropometric data was collected prior to testing and all heights were set in relation to the knee height of the subject, with the exception of the traditional engineering test-rig, which was fixed at 800 mm high. It was determined that there was no significant Body Part Discomfort (BPD) beforehand.

All tests were stopwatch timed, measuring either the completion times for the prescribed tasks or the quantum of task for a given time. The tasks measured hand manipulative tasks, hand ballistic task, pick and place and industrial assembly tasks. All tasks were recognised standard psychomotor tasks except the electric plug-top task, which was developed at the Ergonomics Research Centre at UL.

At the end of each test bank at the prescribed height the subjects filled in a BPD form, adapted from the Corlett and Bishop, 1976[3] models. The outline divided the upper body into eighteen zones, and the BPD for each zone was rated by the subjects on a scale of 0 to 5. With 0 representing no discomfort and a rating of 5 for high/severe discomfort. The diagram represented the body view from the back and eighteen parts were identified, from waist up.

2.4.1 Surrogate Wheelchair User Tests

The Surrogate Wheelchair Users (SWUs) were chosen to approximately match the Wheelchair user cohort. The ages ranged from nineteen to fifty-eight. There were six males and five females. All subjects used the same wheelchair but the footrests were adjusted to suit leg length, so that the subjects seated position was best-fit for the individual. None used armrests. In addition to the anthropometric data gathered from the WUs, the standing stature and standing elbow height of the SWUs was taken. Otherwise the procedure was the same as for the WUs.

3. ANALYSIS OF THE TEST DATA

The test results were analysed in three ways:

- Best-fit, performance and BPD for the WU cohort
- Best-fit, performance and BPD for the SWU cohort
- The comparative analysis of the resulting data for the two cohorts, to determine whether a factors-of-difference existed

3.1 Ergonomic evaluation and comparative analysis of working heights

This section determined the effects of working height on BPD, estimated endurance time, height rating and completion times. Each variable was tested at five levels on the auto-adjustable workbench. The objective of the test was to identify an optimum working height, relative to the user's knee height (KH).

3.2 The effects of working height on BPD

BPD was measured in eighteen body part regions for wheelchair users, after they had completed all seven tasks at each level of working height. Analysis of the test data revealed that KH+100 was identified as resulting in least discomfort among subjects, while KH+0 resulted the greatest discomfort. It also shows that, as the working height deviates upward or downward from KH+100, that BPD increases. A Friedman Test (a non parametric alternative to a within–subject analysis of variance) was used to analyse whether the working height had a significant effect on individual body parts. The results of this significance test for each of the eighteen body parts measured (see Figure 1) are shown in Table 2. The statistically significant parts (<0.05) are underlined.

Table 2. Friedman test results on the effects of working height on BPD – WUs.

Part	Chi square	Sig.	Part	Chi square	Sig.
1	12.233	0.016	**10**	11.688	0.020
2	6.452	0.168	**11**	7.789	0.100
3	17.972	0.001	**12**	10.094	0.069
4	20.058	0.000	**13**	10.359	0.035
5	19.789	0.001	**14**	9.521	0.049
6	12.134	0.016	**15**	8.500	0.075
7	14.261	0.007	**16**	4.388	0.356
8	12.989	0.011	**17**	7.273	0.122
9	12.092	0.017	**18**	11.347	0.023

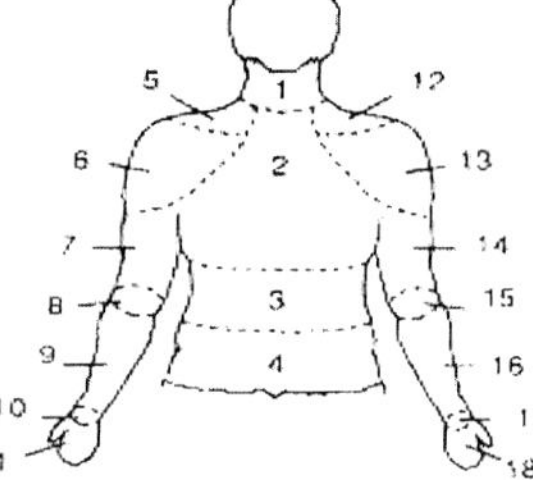

Figure 1 – BPD areas

The results in Table 2 indicate that the working height had a significant effect on twelve of the body parts tested, confirming the importance of best-fit working heights in order to reduce discomfort, and as a result reduce the likelihood of injury, in the long term. Body parts 3, 4 and 5, the waist, lower back and left shoulder were highly significant statistically for WUs.

Table 3. Friedman test results on the effects of working height on BPD – SWUs

Part	Chi square	Sig.	Part	Chi square	Sig.
1	12.672	0.013	**10**	5.644	0.227
2	17.824	0.001	**11**	2.813	0.590
3	21.161	0.000	**12**	13.271	0.010
4	20.275	0.000	**13**	9.932	0.042
5	14.521	0.006	**14**	6.852	0.144
6	9.096	0.590	**15**	1.114	0.892
7	9.442	0.510	**16**	1.429	0.839
8	5.134	0.274	**17**	4.247	0.374
9	1.778	0777	**18**	6.540	0.162

For the SWUs, parts 2, 3, 4 and 5, upper back, waist, lower back and left shoulder had greatest statistical significance. Parts 3, 4, and 5 were common to both groups.

3.3 BPD comparisons of existing workbench to best-fit workbench height

The mean BPD data for working at the existing workbench and working at KH+100mm on new design are contained in Figure 2 below. Discomfort was measured on a zero to five scale on the Y-axis, where zero indicated **'no discomfort'** and five indicated **'severe discomfort'.**

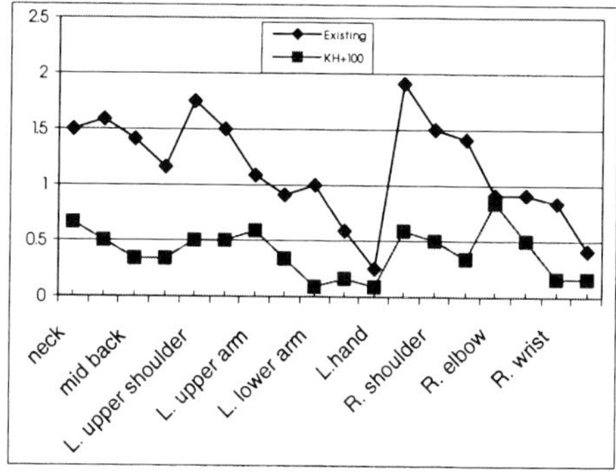

Figure 2. Mean BPD scores for existing workbench and preferred height (KH+100) on test workbench for WU group

The BPD line graphs for the WU and SWU (Figure 3) groups are similar and when subjected to comparative statistical analysis show no significant difference. The preferred height graph is very similar for both groups indicating a highly significant value for the SWU group of, p <0.001 and for the WU group a value of p <0.001. Therefore for both groups the identified, preferred working height of KH+100 significantly contributes to the reduction of BPD. The visual inspection of both graphs, Figures 2 and 3 reinforce this.

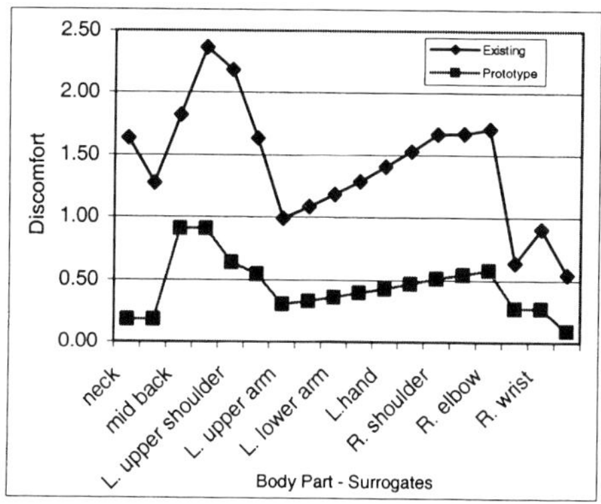

Figure 3. Mean BPD scores for existing workbench and preferred height (KH+100) on test workbench for the SWU group

The results show that there was a significant effect on eight of the body parts tested, most notably all back regions and the shoulders, for the WU group and there were six body parts showing significant discomfort for the SWU group.

Five of the affected body parts are common to both groups. The neck did present a problem for the surrogates, but did not for the WU group. Body part 13 and 14 registered significant discomfort for the WU group but did not for the SWU group. The general discomfort for the total cohort is registered in the shoulders and back. Therefore the preferred height of KH+100 contributes to neutralising body posture for seated workers whether WUs or SWUs.

3.4 The effects of working height on estimated endurance times

Figures 4 and 5 below, indicates the mean estimated endurance times for both groups, while operating at various working heights. On the Y axis below 1 indicates **'less than two hours'**, 2 is **'two to four hours'**, 3 indicates **'four to six hours'** and 4 indicates **'six to eight hours.**

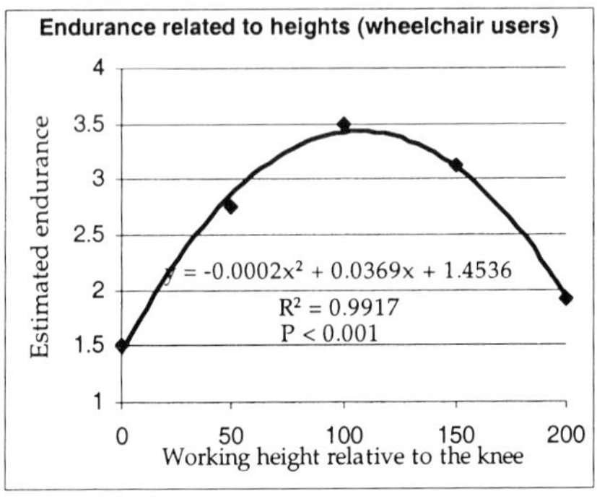

Figure 4. The relationship between bench working-heights and endurance for WUs.

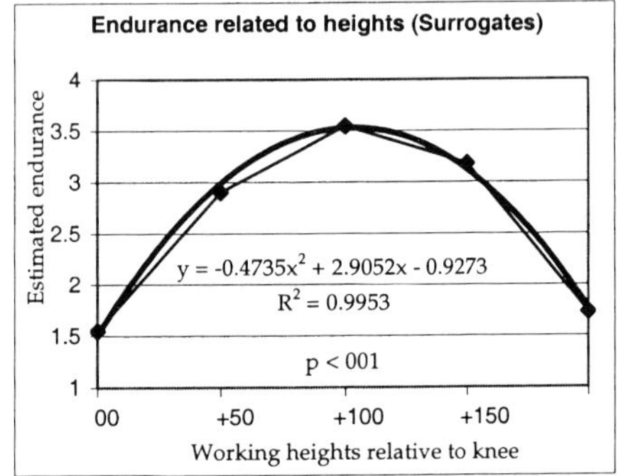

Figure 5. The relationship between bench working-heights and endurance for SWUs

The quadratic equation added to the graphs, may be used to predict estimated endurance times relative to working height above the KH of the wheelchair user, as well as for surrogates. As there was found to be no statistical significance between the endurance for both groups ($p > 0.05$), then it is appropriate to use surrogates to determine endurance. However differences in BPD may need to be considered as this might have a significant affect during prolonged activities. The data collected was based on the test-subjects estimated time that they felt they could spend doing the type of activity associated with the test bank. *A comparative analysis of the curves and their values show that there is no significant factor-of -difference between WUs and SWUs associated with endurance.*

3.5 The effects of working height on completion time

The data revealed that working height does not have a significant effect on completion time for tasks with the exception of the one hole test. There is however a significant improvement on completion times for both the WU and the SWU groups when comparing the existing bench rig with the KH+100 rig. Analysis of WU group data using Friedman test, produced a significance value of $p = 0.010$, and for the SWU group $p = 0.002$. There is therefore a significant improvement in performance for both groups at the preferred bench height, and again the application of any factor-of-difference between the groups was seen as unnecessary.

3.6 Subjective rating of the working heights

Subjects rated the working height upon completion of all seven tasks, at each level. Figure 6 below shows the mean results of these ratings, for the WU group, using error bars with a confidence limit of 95%. Zero on the Y-axis indicates a working height of **'too low'** 5 is **'adequate'** and 10 is **'too high'**.

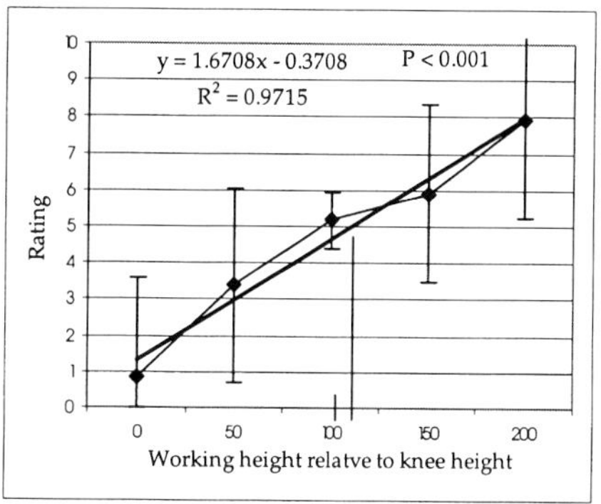

Figure 6. Work heights rating - WU group.

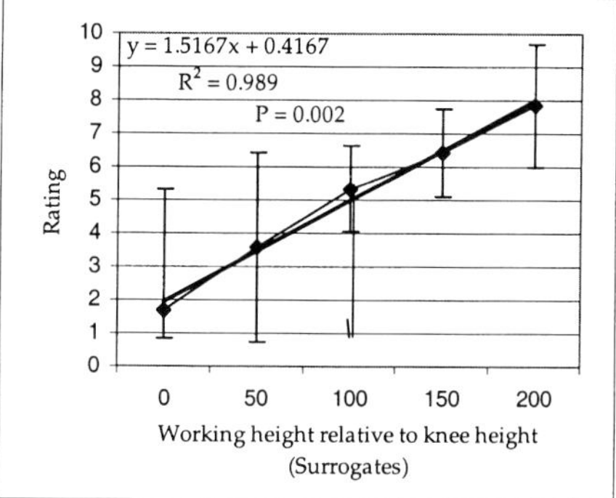

Figure 7. Work heights rating - SWU group.

These results indicated that the working height had a significant effect on the subject's rating ($p<0.001$) for both groups, as seen in Figures 6 and 7. However, there was a significant difference between all others. The results suggest that there is a near perfect correlation between working height, and rating as $r = 0.97$ and 0.99, a positive linear relationship exists. Overall, there is no necessity to apply a factor-of difference for working heights, between the two groups, as the subjective rating was almost identical for both.

4. Existing workbench design vs. preferred working height on test workbench

As an optimum height was identified (KH+100), it was now possible to compare this height with the existing workbench, in order to analyse the effects of workbench design on BPD, completion times and estimated endurance time.

4.1 Estimated endurance times for existing workbench vs. preferred workbench height

The results of the mean estimated endurance time indicate that subjects will work on average two hours (1.75) on the existing workbench, while they will work six hours (3.5) on the new accessible height (100mm above knee level). A Wilcoxon non-parametric test was used to identify whether the populations had the same means. The results ($p=0.002$), indicates that there was a positive increase in endurance, resulting from working at KH+100.

4.2 Productivity comparisons for existing workbench and preferred height workbench

Comparisons were made for the 'existing workbench' scores and the 'best-fit' workbench scores, in order to determine the effects of workbench design on productivity. The results indicated that productivity has increased in all tasks with the largest increase being for the three-pin plug assembly. The results of all tests show that there was a significant difference for all completion times, with the exception of the grooved pegboard, as a result of working on different workbenches for WUs. A similar increase in productivity was established for the SWU group at the KH+100 bench height. The improved productivity for the three pin electric plug was highly significant for both groups.

5. DISCUSSION

The research set out to establish a best-fit datum for working heights for wheelchair users at industrial tasks and to determine if surrogate wheelchair users may also be used in establishing the data. The following conclusions were seen as significant.

- The most appropriate height for wheelchair users (including surrogate wheelchair users) was found to be at Knee Height plus 100 millimetres, i.e. KH+100.
- Working at this height improved productivity and reduced body part discomfort.
- Endurance curves and equations have been devised which will allow the calculation of estimated endurance for a range of heights.
- Moving from the identified optimum in either direction will result in reduced endurance times and increased body part discomfort.
- The comparisons between working at the existing/traditional engineering bench, and the best-fit KH+100 bench produces highly significant statistical evidence of reduced body part discomfort, and improved productivity and work endurance times.
- There was a distinct correlation between the WU group and the SWU group in nearly all facets of the test results, with the exception of some elements of BPD. There was however, a very good match, in the BPD analysis of back and shoulders for both groups.
- There appears to be no ethical reason why surrogate wheelchair users may not be used as test subjects in research relating to bench design to accommodate wheelchair users at industrial tasks. The research also suggests that it is appropriate to mix test groups.

The combined group of WUs and SWUs have not been analysed as a whole. However the findings show that they can be. This will increase the test cohort for the overall research. It will also allow easier access to suitable test subjects. This element of the research has significantly extended ergonomic test methodology and data for wheelchair users at workbenches. In answer to the ethical question of the legitimacy of using surrogate wheelchair users to establish such ergonomic data, the test results speak for themselves – yes, it is ethical.

6. REFERENCES

1. Goldsmith, S. *Universal Design.* Architectural Press, Oxford. Sloan School of Management. (2000)
2. Goldsmith, S. *Designing for the Disabled.* Third edition. Fully Revised, RIBA Publications, London. (1976)
3. Corlett, E.N. and Bishop, R.P. *A technique for assessing postural discomfort. Ergonomics*, 19, 175-182. (1976)
4. Donnelly, A *Statement on the draft Disability Bill*, Ahead Press, Dublin.
5. Drury, C.G. and Coury, B.G. (1982) *A Methodology for Chair Evaluations*, Applied Ergonomics, 13, 195-202. (2002)
6. ISO ISO7250 *Basic human body measurements for technological design.* (1996)
7. Kroemer, K.H.E. and Grandjean, E. *Fitting the Task to the Human: A Textbook of Occupational Ergonomics.* Fifth edition. Taylor and Francis, London. (1997)
8. Nowak, E. *The Role of anthropometry in design of work and life environments of the disabled population.* International Journal of Industrial Ergonomics, 17,113-121. (1996)
9. O'Herlihy, E. and Gaughran W., *Paraplegic trainees and operators in engineering environments.* Proceedings of 2002 ASEE Conference (Montreal), June, Session 2793.
10. Pheasant, S., *Bodyspace: Anthropometry, Ergonomics and the Design of Work.* Second edition. Taylor and Francis, London. (1996)
11. Wilson, J.R. and Corlett, E.N., *Evaluation of Human Work: A Practical Ergonomics Methodology.* Second edition. Taylor and Francis, London. (1995)

Related Topics

Development of concept designs for a disaster relief shelter a student project

P BARBER
School of Art, Design, and Technology, University of Derby, UK

ABSTRACT

The paper discusses the development of concept designs for a disaster relief shelter as a student project. It shows how the concepts where developed and the teaching methods used. The project was done in collaboration with JCB were the initial idea was developed, JCB's aim being to design a workable shelter that could be used for humanitarian aid.

1. INTRODUCTION

The student project detailed in this paper was the development of a concept design for a disaster relief shelter. This project came to the University of Derby from JCB, where several of their employees had identified the need for a redesigned shelter as the existing ones were often nothing more than poorly made tents. The University was approached by JCB to develop some concept designs in order to test their feasibility. It was decided to turn the problem into a student project that students tackled within their Engineering Design module.

The disaster relief shelter offered an interesting and different problem for the students to attempt and also an exercise that could in the future offer some real help to people in situations where their permanent dwelling was not available. The problems faced in the design represented all typical design problems, including dealing with the problems of cost, weight and materials and the trade off required meeting the brief. The project also offered the added benefit of being for an outside client and gave students the experience of working for an external clientele with all the problems and challenges this produces.

2. THE MODULE "ENGINEERING DESIGN"

The project described in this paper is being undertaken as part of the module Engineering Design, the aim of this module is to develop the students' abilities in the application of good engineering principles in order to produce a fully functional design. The module is a 2nd year degree module which is taken by a wide range of students from a variety of programmes including, product design, engineering product design and applied technology. All these programmes are based in the School of Computing and Technology.

The module teaches them the fundamentals of engineering design through a series of lectures and tutorials and a group project which is developed throughout the 12 teaching weeks of the module, on completion of the module the students have to present there work and provide an individual report detailing their contribution to the design. The group project forms the key to the teaching methodology and its requirements form the basis of the underpinning knowledge provided in the lectures and tutorials. Selecting the correct type of project is vital to the success of the module.

The disaster relief shelter project provides all the elements required, as in order to meet the brief a good understanding of engineering materials and principles are required. It is also a different and interesting product which keeps the students' interest and enthusiasm, throughout the module.

3. THE BRIEF

The original idea for the brief was provided by a contact at JCB, they had seen pictures of refugees in Afghanistan and noted the poor state of the shelter provided for them. This had inspired them to see if there was a better way of housing refugees and if it was a feasible project for JCB. At this point JCB did not know how to proceed with the project and it was suggested to them that they might provide it as a student project.

Thus JCB provided the idea and an initial brief, which was to determine the feasibility of such a project including its likely costs, production requirements and possible benefits to JCB. The brief was then refined to make it suitable for the project in the Engineering Design module.

The main aims of the brief were to:-

Develop a "rapid response" basic shelter that can be used for disaster relief and humanitarian aid. The shelter must meet the following requirements:-

- A fully enclosed kit of parts
- Lightweight (2 people able to carry)
- Quick to build
- Require no tools to build
- Able to be parachute "dropped" and air portable
- Durable
- Cheap

These requirements were kept as simple as possible and then the students were encouraged to develop the requirements further in order to produce a specification for the shelter. This work was conducted in their groups with the lecturer answering any questions on behalf of the client. At this point many areas were identified for further investigation including:-

- The size of the shelter
- Possible weather conditions it must survive
- Materials to be used
- What is "cheap"

It was left upto the students to define these areas after the input from the lecturer on weather conditions etc.

4. TEACHING METHODS

The module was delivered using several different methods inorder to provided the students with the learning experience they required. A series of lecturers and tutorials formed the core of the teaching with the lecturers concentrating on providing the students with the required knowledge. These lecturers covered the basics of mechanical design and how it can be applied.

The tutorials were devoted to the development of the designs with the students working in their groups and the lecturer acting as a consultant advising on the progression of their designs. This arrangement meant that the lecturer could monitor the progress of the designs and determine how well the process was progressing.

Several of the tutorial sessions were devoted to specific areas including the investigation of current products and materials properties. These sessions were required to allow the students to relate the theory within the lecturers to practical situations and to experience problems with the current types of shelter for themselves.

The web was used extensively in the teaching of this module, its main use was a resource for the students providing sources of information on the key areas of the products development. The sources of information listed on the website took the form of other websites and links that could be accessed by the students. The website also built up a series of photographs and other images of the students progress, their product evaluation tests and images other useful items. This resource provided a record of how the project was progressing and provided a library of images that could be used in their project reports.

5. PRODUCT INVESTIGATION

The investigation into the current availability of disaster relief shelters and other possible alternatives formed a critical part of the work as few of the students had any real experience of the conditions which could be encountered or the circumstance in which this type of equipment would be deployed. Their investigation took several strands which are detailed in the preceding sections.

5.1 Web Investigation

The use of the internet in determining what was already available and the type of conditions that disaster relief shelters had to function in proved vital as it was able to provide up to date information and first hand accounts. The internet searches done by the students showed that there was a wide range of shelters currently available, but the majority of these tended to be large ex-military type tents, which tend to be heavy and difficult to erect. The web searches did though provided some interesting solutions to the problem in the form of large dome type tents using plastic as a covering and plastic tubes for poles (1).

The results from the students searches along with the lecturers findings were pooled together to provide a list of useful internet sites. This information was then placed on the modules website to allow all the students access to it (2). This proved an effective means of executing a comprehensive web search as it allowed different groups to concentrate on different areas of the design it also meant that all the groups had access to the same information with regard to the current availability of disaster relief shelters and there deployment in the field.

5.2 Tents

It became clear at the outset of the project that the experiences of the students were wide and varied with regard to tents and there operation. It was decided to dedicate a series of tutorials, to the investigation of the practicality of certain types of tent. To do this it was arranged to borrow a selection of tents some of which were provided by students. The students then attempted to erect the tents, noting what problems they had.

This exercise was conducted within one of the universities workshops as this allowed the work to be done in the dry and provided a stable environment in which to determine the difficulty of erecting the various types of tent being tested.

The tents which were tested in this way included a large frame tent and a quick erecting dome tent. The large frame tent was approximately 5m square and consisted of an aluminium frame and canvas covering. It came in 3 bags and was considerably heavy and difficult to transport. In order to determine how difficult it is to erect and the time it takes several student volunteers were selected, these being people who had little or no experience of camping or this type of tent.

These volunteers were initially given no instructions on how to proceed with the erection of the tent. Figure 1 shows them attempting to erect the tent. It soon became clear that without instruction they could not determine how to do it. They were then given instruction on how to erect the tent, with instruction and 4 people working on the tent they managed to erect it in twenty minutes this though did not include the pegging down of the tent which from the lecturers experience can easily take another 20 minutes.

Figure 1

The second tent was a quick erecting dome tent, this was demonstrated by the student who brought it in (Figure 2). This tent was erected in under 5 minutes, without its pegs these take another 5 minutes to put in. This showed how a modern design of tent could be put up in a considerably short space of time. The tent also had a small pack size and a very low weight.

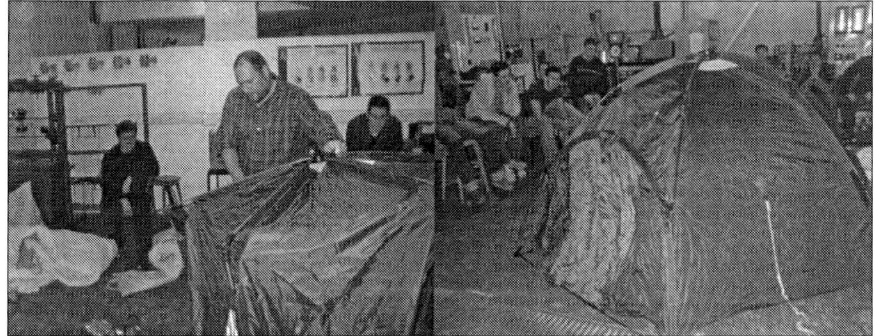

Figure 2

These two practical demonstrations allowed the students to experience what these types of tent were like and how easy it was to put them up. It also showed how much space they enclosed and how usable that space could be. The ideal conditions within the lab meant that the times achieved were the ideals for those particular types of tent.

5.3 Materials

The area of identifying correct materials proved to be a difficult one as the traditional tent making materials and the obvious ones to investigate first. These being canvas and rip stop nylon proved to be prohibitively expensive in both cases and also it was discovered that canvas can double in weight when it is wet.

These discoveries led to the students investigating alternative materials which would fit their briefs for the project. The students identified a number of materials that would meet their specifications, these materials were a selection of sheet plastics including PVC sheet which is cheap and effective, but not as durable as either rip stop nylon or canvas.

5.4 Experiences

The students involved in the project had a surprising low level of experience with regard to camping and outdoor sports etc. This meant that only a small number of them had any real experience of what is was like to be exposed to the elements for a prolonged period of time and the problems that are encountered with traditional and modern tents. This meant that use of groups was vital in disseminating the experience of those who had experienced these conditions and the problems involved in coping with them.

6. THE DEVELOPMENT OF DESIGNS

It became clear early in the project that a traditional tent would not forfill the design requirements and that a different approach to the problem would be required. The initial work of the students focused on attempting to determine what was feasible within the design constraints and how these could be achieved. This work rapidly lead to a selection of different designs being investigated for further evaluation, these being:-

- Yurts:- A development of the traditional yurt as used by Mongolian tribes for centuries. These structures consist of a circular base supported by poles with a shallow domed roof. They are traditionally made of canvas type material. This design can be adapted to use modern materials and others a strong structure with a minimum amount of materials used.
- Domes, Tunnels and other modern tent designs:- The majority of these modern tent designs provide many advantages being both light and easy to erect, but the materials used in them are often expensive and the majority of designs are small in size, having been designed for hiking or expedition use.
- Poly Tunnels:-These are the polyether tunnels used to grow fruit etc in. These tunnels are cheap to manufacture and quick to erect and they provide a high degree of protection. There problem is that they are hot inside as they use clear plastic this could be solved by using a different colour and or grade of plastic.
- Solid Structures:-The concept of a fold out structure was investigated by one group, this proved to have many problems as the folder structure would always be large in size and complex to erect.

These investigations showed that there were only a small number of possible designs that were worth developing further. These being a Dome/Yurt, Tunnel and Poly Tunnels. The groups decided which type of design they felt was most appropriate and from there continued their development towards a completed design.

7. THE CLIENT JCB

Once the initial design ideas had been developed to a reasonable stage and the students had gained an understanding of the problems involved in designing a disaster relief shelter, they were asked to prepare a selection of concept designs to be reviewed by the client. The review was carried out by the client and consisted of the students in their groups presenting their work to the client and receiving feedback from the client.

The review sessions were held in the schools design studio and the students were provided with space to display their designs and given the chance to present their work to the client. The client JCB provided several people including the initial implementer of the project to review the designs and give the students feedback on what they though and how they felt the designs should progress.

This session was extremely useful for the students as it provide feedback on how the designs were progressing from an outside source and exposed them to the realities of designing products for an external client. The client was impressed by the range of ideas and designs that were being developed by the students. The client expressed an interest in many different concepts including a pull out design and several quick assembly designs.

Figure 3 shows the type of displays the students developed for the client which typically included a selection of concept designs and further developments. The boards also identified critical areas for further design work and possible solutions to some of the problems encountered.

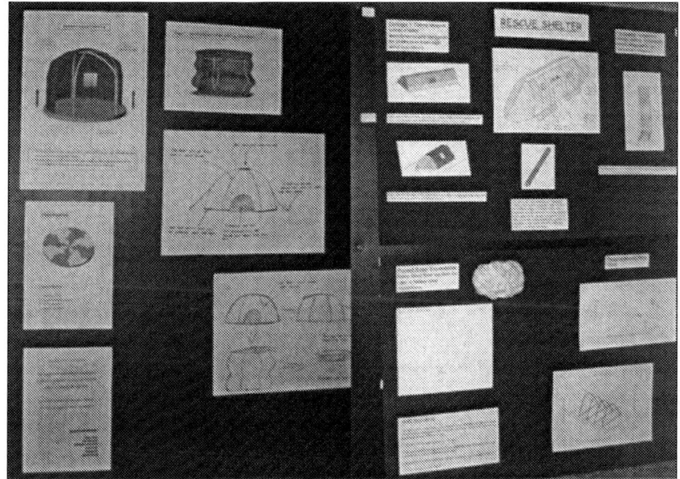

Figure 3

8. DESIGN DEVELOPMENTS

The students assessed the clients feedback on there designs and determined which one they would develop to the final design. At this stage the designs were beginning to converge into two distinct groups, these being designs based round dome / yurt type tents and those based around tunnel type tents. The students in their groups were encouraged to develop their groups own ideas so as not to get a set of identical designs upon completion.

During this stage of the design process it became important for the lecturer to keep the students focused on the original design requirements, as at this point they can become bogged down in the detail of the design and forget the bigger picture. In this case this problem tended to lead to over complex designs as more and more items were added to solve problems.

With this problem controlled the designs progressed well and detailed drawings were developed along with costing's for the design and offer relevant information including materials usage, manufacturing techniques and storage methods. Figure 4 shows an example of the 3D CAD models developed.

The final outputs for the project consisted of a group presentation and an individual report detailing what each student had contributed towards the project. The individual report was required as it provides a clear indication of the work done by each individual student and helps remove the problem of "free loaders" in group work.

The presentations were conducted by the group to the lecturer and representatives from JCB. Each group had 15 minutes to present their design and answer questions on it from the audience. The students performed well at this task presenting their final designs in a clear manner which the JCB representatives were impressed with. It is policy within the school to make the students present their work in many different ways as this provides them with valuable experience in communicating their design ideas in many different formats.

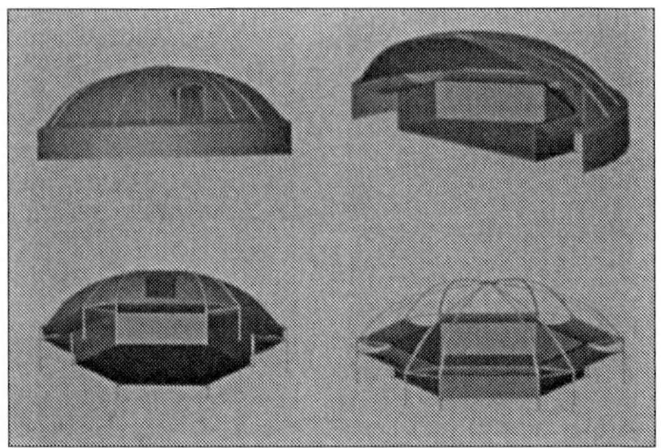

Figure 4

9. CONCLUSIONS

The project provided an interesting and different design problem that none of the students were expecting to tackle at the start of the module. The problem itself provided all the classical design problems as cost, weight, complexity and materials all had to be controlled and trade offs and compromises had to be made inorder to produce a functioning design.

The introduction of an external client in the form of JCB added a new dimension to the project as this was the first time many of the students had worked for an external client. The use of an external client was a very useful tool as it meant that the students had to develop and communicate their design ideas to a different group of people with different objectives and outlooks to the lecturers they were used to developing ideas for. This provided them with valuable experience in dealing with different clients and their aspirations.

Finally the project has helped to further develop a usable disaster relief shelter that can be used all over the world and will to protect people and aid there survival. The eventual aim of the project is to develop a design into a full production version that can be used in this way.

REFERENCES

(1) http://www.shelter-systems.com/
(2) http://depot.derby.ac.uk/student_view/

AKNOWELDGEMENTS

JCB for their cooperation and the initial idea

Copying – a constructive process

C DOWLEN
South Bank University, London, UK

ABSTRACT

Initially the paper looks at ways of avoiding the problems of copying, and then proposes methods where copying can be useful within an educational environment. The first is to use copybook-type processes, where students copy the minutiae in order to develop skills and understanding of the product area. The second is where the essence of copying is used as a base to develop understanding of the work copied with the addition of significant creativity. Copying should not have the bad press that it seems to have developed, but can be used in a positive manner as an educational tool.

1. INTRODUCTION

Copying seems to have a bad press. The point of this paper is to look at the good points of copying rather than the bad points. In some respects it continues the discussions from the E&PDE 2002 paper looking at the topic of sharing (1). The context of the paper is the education of designers and design engineers.

1.1 Educational objectives

To start with it is important when we are looking at educating design students to define the objectives for both the overall courses we run and for each individual element that makes up each course.

Whilst the overall aim of most of our design courses is to produce functioning designers, ready to develop their own work within industry, that cannot be the objective of each individual piece of work. In many instances what is required is to train students in the use of particular techniques and in the development of specific skills. It is inevitable that these specific outcomes should be developed during the early stages of the course, providing foundational techniques on which the student can build and make their later work uniquely their own.

For many of these course elements, a central plank will need to be the process of copying.

2. PROBLEMS – THE BAD PRESS

2.1 Plagiarism

Students' work has to be substantially their own work. We are supposed to be assessing their performance, not someone else's. If students deliberately use someone else's work and pass it off as their own, we are unable to assess that effectively, correctly or reliably. Essentially, students are cheating their way to obtaining an assessment that is not relevant to their own abilities. We need to find ways of reliably assessing students' own abilities and achievements. This is not an option.

2.2 Intellectual property

The purpose of intellectual property was originally not simply to prevent other people using someone's work without permission, but was so that the work could be published and used without the owner fearing that someone else could obtain the benefits from that work being published.

In the days when everything was a one-off, with monks laboriously copying out manuscripts in order to produce texts and each manufactured item made as a one-off, the idea of copying was present, but as it was all done by hand it did not have quite the same import as it does now, where mass publication and copying using mechanical processes are the norm. These provide significant exploitation possibilities, not just from printing but also from software-based publication processes such as the web.

Indeed, design registration defines an industrial design as being designed to be produced using an industrial process. What we are teaching the students to do is to be able to define and determine what is required so that they can be copied as defined.

Patents and design registration documents define how the invention works or what it is. The definition dictates how a copying offence is determined. In the case of a patent, this is to do with contravention of the claims: for design registration, to do with similarity of form.

With copyright and design rights, the system is not pro-active, and the onus is on the holder to prove copying.

3. ADDRESSING THE ISSUES

So how do we determine that copying has taken place?

That doesn't appear to be the real issue. Within education the real issue is how to produce a reliable assessment of student performance.

3.1 How to prevent plagiarism

This is more important than catching it. Assignments can be set so that plagiarism can be minimised, if not avoided. With design work this is relatively easy, in that it tends to be based on a mixture of abilities and skills, usually including an element of presentation work done by the student. Traditional ways of assessing design work include critiques, with students presenting their work orally, and plagiarism is difficult to get away with in this atmosphere. In many cases, such as final year project work, students will in any case each be tackling different briefs: this approach can be carried out at other levels as well.

Group work also tends to make plagiarism difficult, but this has other assessment hazards and can rely on honesty and freedom from vindictive behaviour to obtain a reliable individual assessment.

Developing a culture of honesty can work – but it isn't the whole answer. When students get into industry they will probably have closer supervision and they are hired to get a job done. If they are hired under false pretences they will soon be out of the door.

3.2 How to detect copying

The well-known answer is that students who copy do not know whether they have copied the good bits or the bad bits. You check by seeing that they have copied the bad bits as well and end up with two bits of rubbish. More sophisticated processes, usually in written work, involve searches for similar sections within cohorts and for style change searches. The bottom line is normally that students copy because they can't do the job honestly, so they will be likely to display ignorance and lack of sophisticated effort.

4. METHODS TO USE COPYING USEFULLY

But copying can be more positive than negative.

Essentially, there are two types of method that can utilise copying. In the first instance there are methods where copying is freely used, and where the copies are as exact as possible, and secondly, there are those processes where the essence of copying is used but not the letter. In these what is copied are the good ideas, concepts and parts that are appreciated rather than the letter of the original.

4.1 Methods using exact copies

4.1.1 Copybooks

These were books used for the teaching and learning of the elementary skill of handwriting. Although they are not currently in favour, the idea behind them is a powerful educative one, provided that it is related to the development of specific skills rather than creativity. This type of process produces excellent 'stroke play' and can provide first-hand experience of various design skills (2). It is clearly closed-ended but is useful.

4.1.2 Mastery techniques

Perhaps the best exponents of these are the Japanese. Apparently, according to Richard Seymour, the Japanese have one word that means both to copy and to learn. For them, then, learning is done primarily by copying. This is a simplification of the language. They actually have the concept of mastery. To achieve mastery one studies the master's work in detail and copies it as closely as possible. The idea is that underlying ways of thinking will become clear and the student will start to take on board the master's philosophy and concepts. This has the advantage of a clear goal. What is not so clear until the process is tried is that constant use develops not just a physical understanding, but also seems to develop a philosophical one as well, adding to the mystique.

4.1.3 Apprenticeship process

It is difficult to determine whether this fits into the first category or the second. This is similar to mastery in that the students are apprenticed to the master. They learn the ways of the master in as much detail as possible, so that methods and processes that the master uses are imprinted. The apprentice gains understanding and incorporates the master's expertise into their work, developing under the watchful eye. The process picks up tacit skills and attitudes as well as concrete ones. It works well where craft and skill are the key effectiveness, and can be used in education by placing supervisors in the master's role. In the apprenticeship system proper, however, work is not attributed to the apprentice but the master. That cannot normally be done in an educational context.

4.1.4 Deliberate utilisation of intellectual property

Drawing and presentation techniques can be improved significantly by copying examples from books (such as Dick Powell's Presentation Techniques (3)) and galleries. Products can be modelled or copied in detail to understand how they have been constructed and how they work. This clearly contravenes copyright, but is a useful process.

Patent documents can provide a rich source of material to copy. It is a useful exercise to find loopholes in them, particularly with the claims. These provide opportunities for copying the essence of the invention without infringing the patent. This useful exercise gives significant understanding of the product area and builds up design confidence. Secondly, for a student's major project, the patent holder can be contacted and permission sought to utilise the patent – about 88% of patents aren't incorporated into functioning products (4), (5). The inventor may appreciate contact with someone who might assist them with the exploitation of their invention. Student projects rarely result in commercially viable products, so only in exceptional circumstances would they result in a valuable infringement.

4.2 Open-ended copying processes

These are ones where only the essence of something is copied, rather than the minutiae. The purpose behind these is to develop discernment and understanding as well as creativity.

4.2.1 Trawling

This involves copying, but not simply that. A description can be found in Petty's book "How to be better at... creativity" (6). The idea is to locate an admired piece of relevant work. The points of admiration are determined. These are incorporated into a piece of work, but not the rest. The process was also described in E&PDE 2002 (1).

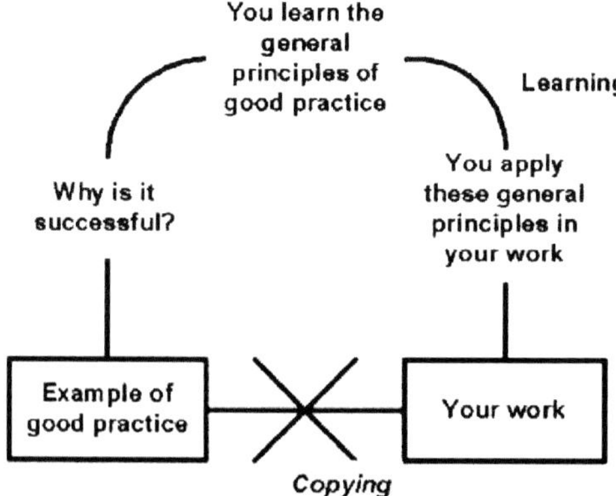

Figure 1: Trawling (from Petty (6))

4.2.2 Evolutionary techniques

There is much talk about genetic algorithms to develop improved products. Whether the evolutionary processes are carried out rigorously or not, the outputs are improved products. Improvement implies that there must be an original to improve. Thus, the starting point for must be something existing, preferably state-of-the-art. This implies that the process starts off by copying.

In any product area such processes are always taking place in an ad-hoc and unstructured manner. Frank Barron says: "New forms do not come from nothing, not for us humans at any rate; they come from prior forms, through mutations, whether unsought or invited. In a fundamental sense, there are no theories of creation; there are only accounts of the development of new forms from earlier forms" (7). He unpicks creativity doctrine and process as being of significant imitation and development. The product area develops as products change: it is unlikely that any product is produced without reference to existing products; if it was it would be unlikely that the market would readily adopt it. In practice, there is a gradual shift after an initial unstructured phase that establishes a product paradigm (8) (9).

5. EXAMPLES

This section will look at examples from teaching. The first set deals with copybook-like situations when training was high on the menu, and the second where a more creative and original approach was taken.

5.1 Training situations and exercises

In these examples the major emphasis is on students developing techniques and understanding rather than developing their own ideas. The issues are not those of copying, but of developing skills and of understanding, Students do not get assessed on their ability to produce original work, but on their ability to perform the skills effectively.

5.1.1 Manufacturing exercises from the first year

At South Bank design students have a manufacturing programme when they start. This involves them in manufacturing basic items such as a plumb bob and a small sheet tinplate box. In these exercises they copy designs using given drawings.

Also in the first year they have an assignment to model a small existing product. They are led through a number of modelling techniques and this results in their developing specific modelling expertise that is difficult to achieve if they have a freer hand. They copy, as exactly as possible, the external form of the product; in this case a Braun alarm clock.

In spite of the copying brief, it is still relatively easy to determine an assessment of each student's abilities – this relates not to how much of the work is original, but in this case, how close the copy is to the original product: almost the inverse of an assignment on the development of creativity.

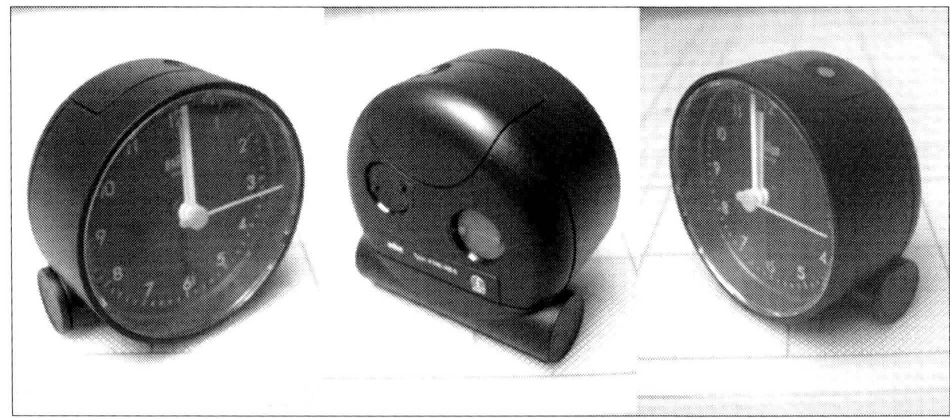

Figure 2: Braun alarm clock

Figure 3: Alarm clock models

5.1.2 Manufacturing drawings assignments

At the same time as carrying out this modelling exercise, students are doing two exercises to develop their engineering drawing skills. They need to develop these first not by drawing the products and components that they will be designing, but by taking other components and assemblies and producing drawings of these such that they can be manufactured. These are either injection moulded or (in a few instances) die cast components, so they are not always the easiest of components to draw. The students need to go through the mental process of copying the component and working out exactly what makes it tick and how it was made so that they can do what is effectively a re-engineering process on it, although in this case the re-engineered components are never actually made. Like many of these exercises, this is designed to develop a certain number of tacit skills in the students: the simple instructions do not talk about moulding design but simply about drawing the component that has been provided.

Here copying is used as a method, but because all the students have different components to draw, they are not able to copy each other's work, although they are able to start off by pooling thoughts such as how to start looking at the component and how to do such basic tasks as incorporate a drawing block.

5.2 Examples of copying aspects of work

In these examples a more original and creative approach is required. But students are still required to carry out some element of copying. This set comprises three different pieces of work. The first example is of a very successful Design Heroes assignment, where students locate a past designer that they appreciate, study their work and propose a way of imposing their ways of working onto a product that their hero never worked on. The second example is in a similar vein, and is the addition of historical styles to products in an artificial manner.

5.2.1 Design heroes assignment

The process of this is very much related to the trawling method, and was construed as a sharing process as well as a copying one, so was mentioned in the E&PDE 2002 paper (1).

It is about developing and understanding style. Students research and locate a designer whose work they admire. Each student is required to select a different designer, therefore cutting down the possibilities that during the research they will simply be cutting and pasting huge chunks of data from an Internet search.

They need to distil the essence of that designer's style, and then apply it to a product that is not one that they actually designed. This means that they are trying to separate out those elements that they admire from the overall nature of the work and understand the nuances of form and (perhaps) function in order to essentially copy the essence of the work.

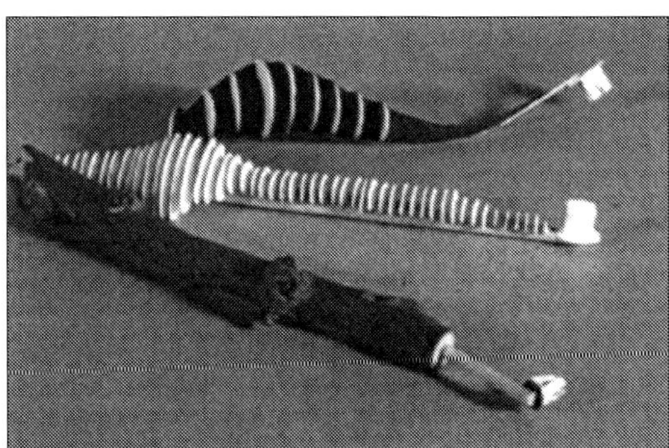

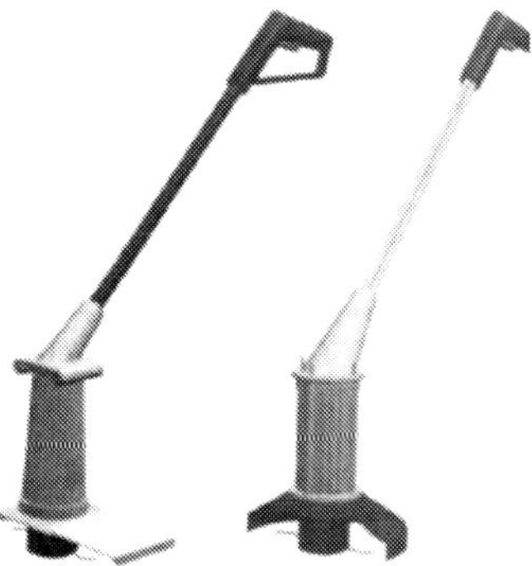

Figure 4: Design Heroes and Style Assignments:
Andrea Branzi toothbrushes and classical and gothic grass trimmers

Normally the designers have a historical context, and students may decide to select designers such as Raymond Loewy, Clarice Cliff, Christopher Dresser, Adrian Branzi, Philippe Starck, Terence Conran, and so on. Or they may be architects such as Frank Lloyd Wright, Le Corbusier or Eero Saarinen: only occasionally did these apply their work to products other than buildings.

The exercise has been a particularly successful one, as students not only learn about design history and start to evaluate products from a stylistic and aesthetic perspective, but they also incorporate this evaluation into a creative endeavour and build on it. However, the designs are not expected to necessarily relate to the state of the art in the product area they have chosen, tending to be somewhat quirky examples and more for use as educational examples.

5.2.2 Style development assignment

In a similar way, students have been asked to investigate specific, historical, named styles and incorporate those styles into somewhat unlikely products such as electrical garden tools and electronic equipment. This results in significant understanding of the essence of the style together with non-superficial understanding of the product function and construction.

6. CONCLUSIONS

This is about useful copying. It sets out the issues but goes beyond them to concentrate on the usefulness of the concept within the product design environment. Copying becomes a positive design process – if the pitfalls can be avoided effectively.

REFERENCES

(1). Dowlen. "*The Place of Sharing Experience in Engineering Design in Teaching*". In *E&PDE 2002*. Coventry: MEP, 2002.

(2). Dowlen, C.M.C. "*Developing the Creative Environment for Engineering Design*". In *Engineering Design and Creativity*. Pilsen: Heurista, 1995.

(3). Powell, D., "*Presentation Techniques*", Macdonald: London, 1990.

(4). Baxter, M., "*Product Design: Practical methods for the systematic development of new products*", Chapman & Hall 1995.

(5). Hollins, W.J., & Pugh, S, "*Successful Product Design*", Butterworth: London, 1990.

(6). Petty, G., "*How to be Better at... Creativity*", Kogan Page: London, 1997.

(7). Barron, F., "Putting Creativity to Work", in "*The Nature of Creativity*", R.J. Sternberg, Editor, Cambridge University Press: Cambridge, 1988.

(8). Dowlen, C.M.C. "*Development of Design Paradigms*". In *International Conference on Engineering Design*. Munich, Germany: Technical University, Munich, 1999.

(9). Dowlen, C.M.C. "*The Evolution of the car: An investigation into product history. Similarities, contrasts and questions.*" In *Design and Nature*. Udine, Italy: Wessex Institute of Technology, 2002.

Citius, Altius, Fortius – integrating competitive training principles into the designers world

B T J DYER
School of Design, Engineering, and Computing, Bournemouth University, UK

ABSTRACT

Design in its current state is constantly striving to get the best results within the confines of a limited resource – time, cost and ability. As a result, designers are now under increased pressure to work faster and smarter within a competitive environment.

However, whilst design education teaches a student design methods and tools, we do not improve the training of the mentality of the designer despite this being one of the main driving motivators of any given action. This paper sources the techniques used under the greatest of human stress i.e. in competitive endeavors, and then adapts these techniques to improve the efficiency of the designer's process before, during, and then after in the form of a design methods model.

1. INTRODUCTION

Films such as Wall Street highlighted the 'greed is good' philosophy apparent in some corporate environments during the 1980s. This focused, almost selfish demographic caused stress, animosity, and then ultimately resentment from other members of staff from outside or within the corporate environment. Long term this leads to *Freudian Death Theory* or overly aggressive behavior in business and life. Later in the career of the individual, *Designer Burnout* (3) is often experienced and this can contribute to a shortfall in both effective and experienced designers.

However, it has been proven in other aspects of life other than business that competitive instinct can be harnessed and employed as a positive practice. This in turn can improve ability without creating direct negative perception from peers and colleagues. This is *non-confrontational* competition.

It has been researched and discussed that sport builds character (2) but it wasn't until the 1950's that researchers began to seriously test the validity of this (8). These studies culminated in socialisation snapshots (2) whereby certain characteristics were noted.

'Performance is a direct function of drive' (7)

Success is defined as 'favourable accomplishment, attainment, or outcome' (4)

Success is evident in its purest form in competitive sport. Sport whilst in essence an extra curricular activity now produces professional athletes whose demands on both their physical and mental ability is designed by definition to determine those limits and to obtain the best ultimate result. Two examples of these are:

'Winning is 90% mental' – (7).

...of 1500m running – 'of twelve athletes on a start line, six hope they are going to win, four think they're going to win, and two know they are going to win. The winner will be one of those two' (9).

A sports analogy has been used successfully as a basis for performance enhancement in disciplines other than design. Professor Keefer has used 7 key characteristics of sport ability to improve the mindset of writing and theatre student's (6). This is grounded by using sports methodology and mental conditioning as a basis for improved ability in creative endeavours. The 7 characteristics are Strength, Endurance, Coordination, Focus, Flexibility, Speed, and Posture. This has led to the adoption of the 'Brain Gymnasium' taught at New York State University taking previously untrained students and developing them in creative arts.

This paper takes standard sports psychology thought processes (7), integrates it with taught design methodology (11), and identifies this in the form of a model to improve designer's output known for reference here as the DPM (The Design Performance Model). This aims to help a designer to retain a positive stance in the eyes of the designer's peers whilst improving personal ability.

2. INTEGRATING COMPETITIVE ABILITY WITH THE DESIGN CORE

One of the more modern and commonly practiced design methods of process structure is Pugh's Design Core (11). Pugh has long been associated with identifying design methods (10) to educate and improve the ability of both students and designers (see Figure 1). However Pugh deals with design methods as 'tools' that are available to use as part of his structured process. These tools do not always train the designer at the 'front end' preceding their design process. In other words 'how well a designer uses these tools'.

MARKET > SPECIFICATION > CONCEPT DESIGN >

> DETAIL DESIGN > MANUFACTURE > SELL

Figure 1. The Design Core.

Pugh's methods follow the basic process flow for a given action shown in Figure 2.

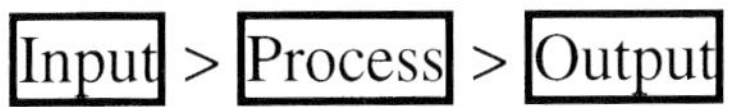

Figure 2.

However to better explain the performance analogy this is modified as in Figure 3.

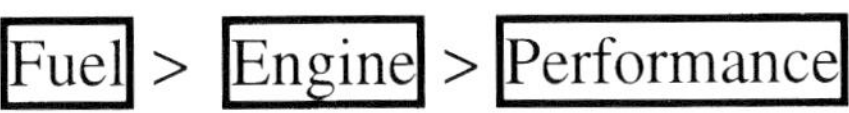

Figure 3.

Pugh's Design Core could be considered here as part of the 'Engine' element. The DPM model described in section 3 begins under the 'Fuel' element and in addition then adds 'muscle' around the design core 'strengthening' the process (Figure 3). In other words, the competitive techniques employed in the DPM are the 'fuel' for the design act. A lack of fuel means the 'engine' (the design process) will be starved in terms of output or quality. The outcome of this is the 'performance'.

3. THE DPM MODEL

The DPM model is generated from standardised sports psychology terminology (7)(8)(12) and is made up of 4 key stages to provide the 'fuel' element (see Figure 3) for the design act at any part of the design core. Each stage below cannot begin until the stage proceeding it is completed. The following sections describe each one of the boxes below in more detail and the techniques thereof.

Figure 4. The DPM Model

3.1 Preparation – DPM Stage 1

The preparation phase is made up of 2 techniques. These are *Character Awareness* and *Periodisation*. These are the foundation for all subsequent actions in the model.

3.1.1 *Character Awareness*

Dr's Ogilvie and Tutko (7) recognised 11 key qualities that are deemed essential for a successful athlete. Knowing and employing these abilities artificially already increases ability in a required activity even if they do not naturally occur. These are:

- Self-motivation, drive, aggressiveness, determination, leadership, self-confidence, emotional control, mental toughness, coachability, and conscientiousness.

Advantages: The amounts of each of these requirements will vary from person to person. There is also the industry held stereotype that younger designers are a better creative resource. By identifying and adopting the above characteristics, allows a reduction in burnout, career plateauing, and stress in more experienced designer's (3).

Disadvantages: Characteristics cannot be employed artificially if these are not part of the designer's psychological profile. However these can be stimulated artificially to a certain extent. Example of these include:

e.g. Performance related promotion/pay, appraisal & feedback, peer related praise, psychometric testing at point of employment.

3.1.2 *Periodisation*

Endurance is achieved by a method known as *Periodisation* (1). The outcome of this practise in modern thinking is the *performance pyramid*. This pyramid model (12) (Figure 5) was created to show how an athlete can develop and then achieve a key performance at a later date without fatiguing both mentally and physically. Designers cannot work at their highest levels of ability permanently. As a result *pace* is required (as is adequate mental rest).

Endurance is created in three key stages. These are *base, intervals, and speed* (12). This is adapted from Bompa's directives and theories in attaining performance (1) and led to the performance pyramid model. This process results in a 'peak' indicated in Figure 5. The resultant peaks intention is an optimum result.

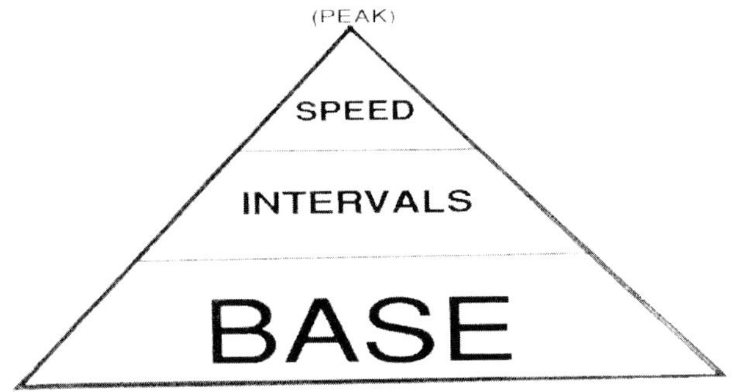

Figure 5. The Performance Pyramid.

The pyramid has 3 key stages and these can be adopted for the designer.
Base is generated during low stress projects or basic tasks that are numerous and are educational for the designer. This base is the foundation for high levels of performance later in their career. These comprise simple one off tasks and could include the generation of a rendering, assisting in CAD drawings, taking part in team activities.

Intervals are whereby greater constraints are placed upon the designer in terms of resource and responsibility. These are variable duration's of design actions and are followed by moments of recovery. Examples of these are short projects whereby the designer is not the project manager but is a contributing factor.

Speed is whereby maximum stress is endured for a short sharp action. This is the period where the best result is needed to be achieved. This could be a key presentation or a re-design of an existing component quickly.

To put this further into the designer's context consult Figure 6. The bigger the base the stronger and larger the pyramid and thus the better the result. A new or junior designer begins with constructing base and progresses inevitably to peak at some later point in their career i.e. promotion or on a key project. It is worth noting that a peak will only last so long and this process is cyclic. As a result, this base will need to be generated again to avoid fatigue or burnout. In working practise this is a project change, staff development or a promotion, depending on the duration of the pyramid. This duration could be monthly, annual, or over any time period during the full career of the designer. This answers some of the concerns regarding long term staleness and fatigue raised in Eckert's (et al) paper on burnout (3).

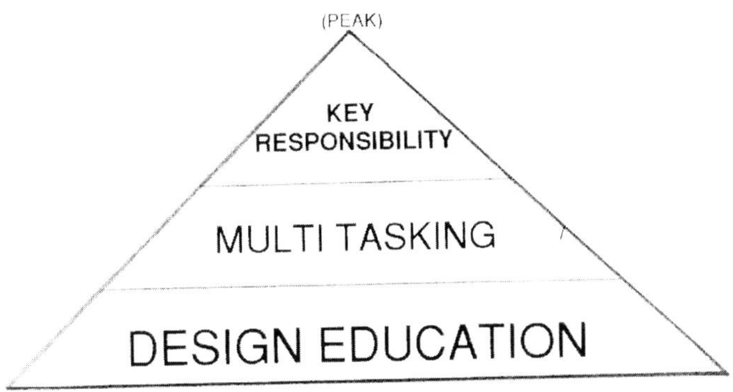

Figure 6. The Performance Pyramid – Modified.

Advantages: The designer engaging in this cyclic method prevents or delays burnout by having a constant refreshment of knowledge avoiding some of the pitfalls of *designer burnout* (3).

3.2 Arousal & Priming – DPM Stage 2

The second stage of the DPM is to make the designer alert and ready to achieve. Arousal in this application is sometimes regarded as Pre-*Anxiety*, *Arousal Theory* or what's sometimes specified as *Drive Theory* (8). These theories are the point directly before a person begins to engage in an activity. Only a certain amount of arousal is ideal. Too much and the feeling then becomes one of nervousness and stress. Given the assumption that most working activities are not new, hopefully this should not be a factor however added responsibility, critical fear of failure, or a definitive goal can cause uncontrolled arousal.

Techniques:

The six guidelines below are specified to improve arousal (8). Again documentation will be needed initially to aid permanent retention. These guidelines are:

3.2.1 *Focus on what you can control.* – Avoid 'what ifs' or 'I hopes'.

3.2.2 *Think practise.* - If a new skill is utilised or an older one being used (such as rendering, presentations or airbrushing for example) consider each attempt as a practise. This will relax the body's motor functions and will increase the chances of first time success.

3.2.3 *Know the worst case scenario.* - What is the worst thing that could happen ? Anticipate failure in staff, the team, or your technique. Know your client. What do they want to see ? What is the brief.

3.2.4 *Keep active.* - Constant activity is proven to reduce anxiety. Less time to worry means less time to think about potential failure.

3.2.5 *Use pre packaged interventions and cognitive strategies.* - This subject is far more detailed than the limits of this paper however this basically covers different mental 'games' and thought processes. For example reciting a rhyme, humming a song, or employing 'visualisation' of a scenario and its success. Other researched models include Smith's (8) Anxiety Management Training, Suinn's (8) Visuo-Motor Behavioral Reversal and Meichenbaum's (8) Stress Inoculation Training.

3.2.6 *Develop a mental plan.* This can be a flow chart or a list of project requirements although in a designers situation the substitution of a Gantt chart or a physical timeplan will probably be more effective to himself and to the connecting team.

3.3 Assault – DPM Stage 3

Despite the constriction of time, the human body responds better to an even pace with little changes in stress and work expectation. A common problem is the designer generating fifty ideas in the first hour of design conception but then nothing for the rest of day resulting in a panic or query at the end. The human body does react well to changes in demands. A constant demand uses less energy and therefore generates less fatigue. The outcome is that mentally the designer will last longer. This is commonly known as *Pace Awareness*.

Technique: A methodical collation of ideas and repetition of research, brainstorming, and focus groups can be employed to continually stimulate creativity. In this case, time constrained brainstorming using a key word from the specification or focus groups with different members.

3.4 Evaluation– DPM Stage 4

The evaluation stage commences after the design act has been completed. Currently this is measured by a job being completed on time, a product working well, and rewards (both financially and by reputation).

Technique: How it is controlled in a competitive environment is to find an ability to 'measure' one particular performance over another. An existing educational practise is to encourage students to use a *logbook* or *experience diary* (5) but this idea is expanded in it's content. It is needed to ascertain not only *what was achieved* but in addition *under what conditions* it was achieved. This improves the chances of success being replicated. The information that should be noted in addition to the activity are factors such as:

Activity duration, time of day, environment, diet, previous sleep pattern, fitness, and personal circumstances/ stress.

Advantages: Records of success can be viewed by the individual. From this it will become apparent the best methods and conditions to repeat project success. Plus, these records can be transferred to other members of staff for objective review and education.

4. CONCLUSIONS

The DPM model uses commonly practised techniques developed from modern sports psychological thinking. Keefer's adaption of sports terminology has shown the benefits of crossover to other creative endeavours. Avoiding burnout has seen the recommendation of refreshing both knowledge whilst developing experience, and this model can contribute to that (on condition both the designer and the rest of the management strućture adopt it).
However, experimentation is required to see the results of the techniques described in this paper. Some techniques work for some but not for all. Practise is required so that the techniques become instinctive in the same way that design as a process becomes natural during undergraduate education.

The results in their sporting context are continually highlighted as performances improve and records are broken. The knowledge employed in university and sports institutes in isolation over the last few decades have now begun to filter down into public awareness and as such can be utilised effectively.

REFERENCES

(1) Bompa T. *'Periodisation Training For Sport'*, Human Kinetics, 1997.
(2) Coakley Jay,J. *'Sport In Society'*, Irwin McGraw Hill, 1998.
(3) Eckert C, Stacey M, & Wiley J.*'Expertise & Designer Burnout'*. Proceedings of ICED Conference, 1999.
(4) *'English Dictionary'*, Collins Gem, 1998.
(5) Jordan, Patrick W. *'Designing Pleasurable Products'*, 2000.
(6) Keefer Evergreen J. *'The Brain Gymnasium'*. NYSU Writing Workshop 2, 1998.
(7) Miller, Brian. *'Gold Minds'*, 1997.
(8) Morris, Tony & Summers, Jeff. *'Sports Psychology'*, Jacaranda Wiley Ltd. 1995.
(9) Pickering, David. *'Cassell's Sports Quotations'*, 2000.
(10) Pugh, Stuart. *'Creating Innovative Products Using Total Design'*. Addison Wesley Ltd. 1996.
(11) Pugh, Stuart. *'Total Design'*, Addison Wesley Ltd. 1990.
(12) Trew,Steve. *'Triathlon: A Training Manual'*, 2001.

Less without loss – connecting producer responsibility with product design in UK SMEs

R HOLDWAY
Department of Industrial Design Engineering, Royal College of Art, and Giraffe Innovation Limited, Brighton, UK

ABSTRACT

The paper derives from two extensive programmes of work. Firstly ongoing research and consultancy work by the author for Envirowise, a UK DTI/DEFRA government programme dedicated to improved environmental practices. Secondly, work by the author on the Good Design Practice Project, a collaborative project between the Royal College of Art and the University of Cambridge.

The paper looks at some of the key organisational aspects of Design For Environment (DfE)

It considers the context, key barriers, potential solutions and success factors. The paper draws upon the authors work in 14 action workshops, with both small and large companies. The paper illustrates how a proactive response to producer responsibility legislation can enable companies to not only minimise the environmental impact of their products, but also achieve significant cost savings – LESS WITHOUT LOSS. (1)

1. INTRODUCTION

The impact of design and management decisions can be seen in the waste filling the world's landfill. The UK alone sends a million tonnes of electrical and electronic waste to landfill each year. This is set to double to more than 2 million tonnes by the year 2010. (2) The equivalent figure for Europe are 6 Million tonnes and 12 Million. Electronics waste (defined as products that depend on electric currents or electromagnetic fields in order to work properly) is the fastest growing waste stream currently growing at around 8% per year. (3)

Increasingly, legislation seeks to compel producer responsibility for all products, and therefore is pushing companies towards increased recovery and recycling of all materials. This is part of grander ambition for 'zero landfill'. This cannot be achieved unless companies redesign their products to minimize life cycle impacts – particularly at the end of life.

This paper seeks to identify the essential principles that minimise their end-of-life impact. An emphasis is placed here on action workshops which address legislative compliance. But more than this – the workshops induce an attitude and behaviour within companies that is BEYOND COMPLIANCE. Enlightened management can address design for environment (DfE) as an opportunity for innovation and differentiation, rather than an unwelcome constraint. In this perspective DfE is central to the design process, and early decision making. DfE principles are thereby embodied in the product outcome.

2. DEFINITIONS

Sustainable Design can be defined as "finding new ways of balancing (present and future) human kinds need for products and services and systems, with the needs of the Environment and society as a whole". (4) The work discussed in this paper deals with the environmental pillar of the triple bottom line (economic, social and environmental) of sustainable development. However, it recognizes that actions to meet goals in any one of these spheres affects the others.

Our workshop focus is to target design solutions – either at the end of their life, or new designs in the process of development. From that point we encourage companies to work back to commitments and rational scenarios within DfE principles. This method from systems thinking is sometimes called 'Mode 2 Robustness Analysis'. See Rosenhead and Mingers 2001 (5)

For the purposes of this paper we will use Design for Environment (DfE) as shorthand for the main theme of the action workshops. Design for environment (DfE) can be defined as- a process to improve the environmental impact of a product by assessing all of the materials and component parts, functionality, and use, across the whole supply chain to end of life. (6]) However, it is also important to acknowledge the need of companies to achieve business success. Therefore a rephrased definition is proposed that prioritizes economic gains in the context of environmental gains: *DfE is the development of innovative, efficient products-while both satisfying commercial criteria such as cost, quality, appearance; and complying with environmental issues over the entire product life cycle.*

The economic argument for undertaking DfE is consistent with a 1995 survey (6) that demonstrated 21% of SMEs perceived costs savings as a benefit of DfE. In the same study 73% of SMEs stated cost savings as the key factor in motivating SMEs to adopt environmental policy.
This is consistent with the experience of Envirowise DesignTrack workshops. One manager commented:
"The effects of that amount of money saved over that short period of time, really hit some nerves. It made us aware of what we could do if we wanted to." [Company Manager]

2.1 Life cycle

The concept of the product life cycle is key when thinking about DfE. The life cycle of a product is the stages that a product travels through from materials extraction and processing, manufacture, use and finally disposal, reuse, and remanufacture. An extended reuse of products and materials is sometimes referred to as 'cradle to cradle' (7) (8). It is crucial to consider the environmental impact of any design at the early stages of the product development process. The decisions made in the early stages of the design process account for over 80% of the environmental impacts over the product lifecycle. (9) By addressing the environmental aspects throughout the complete product or service life cycle, environmental impact can be greatly reduced.

3. GREEN TAPE

Concern about sustainability operates at two levels: as an anticipatory response to long-term global crisis, and as an immediate response to legislation. Most environmental thinking is reactive and deals with 'end of life' impacts. The enactment of new environmental laws is rising rapidly, with 300 new EU Directives and Regulations being introduced on top of existing national ones. (10)

3.1 Producer responsibility

Producer Responsibility is a business-led approach to reducing the impact of waste products in the environment, in particular by reducing waste, and by increasing reuse, recovery and recycling rates. It is intended that those who manufacture, distribute or sell products or materials should take a greater share of the responsibility for what happens to those products or materials when they reach the end of their lives.

There are two European Union Producer Responsibility directives that are currently relevant to the recycling and the ability to recycle products (11). These are Waste Electrical and Electronic Equipment (WEEE), and "The Restriction of the use of certain Hazardous Substances in electrical and electronic equipment" (RoHS).

The first directive sets recycling/recovery targets for products. All incurred costs, from the collection points to the environmentally sound treatment, re-use and recycling, will, ultimately, be covered by producers for their own products. The second directive bans four heavy metals (lead, cadmium, mercury and hexavalent chromium) and the brominated flame-retardants PBB and PBDE from 1st July 2006.

These directives were ratified last October (2002) by the EU, with the aim to transpose them into UK law by August 2004. The national recycling and recovery targets have to be met by 31st December 2006. This requires manufacturers, suppliers and designers to consider the implication of their product designs from the outset, to meet their obligation to EC directives.

This legislation represents the need for a paradigm shift in a company's approach to design – that of 'whole life' thinking factored into the design process from the outset. The emphasis has shifted from a vague intention that products ***can be*** recycled to the obligation that ***it is*** recycled. While 'whole of life' thinking is not a new concept, there is much evidence from our work to suggest that companies, and policy makers have not yet realized the practical implications. Most UK companies are unaware of the impact the legislation will have on the design of products.

Other EU policy tools are under review. One is IPP – Integrated Product Policy, a future policy based on the environmental impact of the whole product lifecycle. There is also a new discussion document (published November 2002) "Eco design of End Use Product (EUP). This further defines the need for specific lifecycle thinking in the product development process. For example, companies may be compelled to construct '*environmental indicators*' on each product throughout its lifecycle. This is likely to be in the form of Life Cycle impact data.

4. INDICATIVE PROBLEMS - FRIDGE RECYCLING EXAMPLE

The problem encountered in recycling fridges (and subsequent fridge mountain) is an example of the downstream problems weak design decisions can engender. The fridge problem in the UK has been addressed with expensive reprocessing machines. One reprocessing company, EMR processes over 70 fridges an hour. A recycling rate of 97% is achieved for steel, copper, aluminum and polyurethane powder. CFC's are removed from coolant and insulation foam. (12) However, there are as many as 5 variants of plastic used to produce fridges. All the plastic from these products currently goes to landfill, as a method of separation has yet to be devised.

5. GOOD DESIGN PRACTICE

Design plays a central role in minimizing a product's impact on the environment. Over the last three years the author at the Royal College of Art (13) in collaboration with University of Cambridge (14) has been working with Small & Medium Sized Enterprises (SMEs) to improve their use of design in the development of new products.

Four discrete issues emerged from our work: *functional orientation*, new product development *(NPD) process effectiveness*, *timing of industrial design input* and *quality of execution of core business processes*

In each area, design activities could be deployed to enhance performance. (15) Nearly every company performed badly when assessed against DfE as a early design concern. This indicates a significant deficit in companies' awareness of the issues. Moreover, the absence of knowledge on key legislative barriers undermines NPD process effectiveness.

5.1 The knowing doing gap

A key issue emerging from our work is the inability of many companies to turn 'awareness' of good business and design practices into 'action'. This gap is termed the 'knowing-doing problem'. *The challenge of turning knowledge about how to enhance organisational performance into actions consistent with that knowledge"* (16).

In a recent survey (17) it was shown that while many companies believe sustainable development to be important, few had taken any substantial implementation steps.
When asked where they potentially could make most progress towards sustainable development more than half of them answered that *product design* was the most promising area for performance improvement.

Even where DfE is an accepted approach, companies tend to embody its principles in unstructured and ad-hoc ways, without proper integration across the organization.
A potential barrier to the incorporation of DfE principles is the awareness and expertise of designers. At present, anecdotal evidence (18) suggests industrial designers know very little about the issues relating to sustainable design and DfE. This is consistent with our direct experience with companies.

6. DESIGNTRACK WORKSHOPS - Practical Environmental Advice for Business

Envirowise is jointly funded by the Department of Trade and Industry (DTI), the Department for Environment, Food and Rural Affairs (DEFRA), and devolved administrations of Scotland, Northern Ireland and Wales. The programme is managed by AEA Technology Plc and the National Physics Laboratory (NPL). Envirowise is dedicated to helping UK companies be more profitable and increase national competitiveness through improved environmental practices. All advice is provided free of charge.

6.1 Designtrack workshop

Design Track is a free and confidential service from Envirowise that focuses on reducing the environmental impact of a product over its lifecycle. The result of a DesignTrack visit is often a product that is easier and cheaper to make and use; and has reduced environmental impact.

DesignTrack focuses on UK SMEs with less than 250 employees. The aim is to enable candidate companies to adjust their practices to the demands of producer responsibility legislation, namely the WEEE, RoHS and End of Life Vehicle (EoLV) Directives.

While legislative compliance is central to this project there are many collateral benefits that can improve competitive advantage:

- Improved product function and quality, and longer product design life
- Reduction in assembly time and lower production costs
- Improved customer/supplier relationship
- Continued compliance with legislation
- Easier equipment disassembly for recovery and recycling
- Improved company image through demonstrating commitment to environmental improvement

In 2002 Envirowise performed a pilot test of 10 Designtrack visits. Due to the successful outcome of these visits, a further 40 have been targeted between October 2002 and April 2003. The author has extrapolated lessons from his original involvement in the pilot scheme and his own on-going work in phase 2. The text is based on 14 workshops.

6.2 Workshop Process

The aim of the DesignTrack workshop is to help companies make cost savings and environmental improvements through design improvements for a particular product or product series. At the most basic level the workshop seeks to raise awareness and change behavior of product developers at the upstream concept development stage, as well as at the specifying and detail design stage. The nature of the workshops is developmental. Participants explore ideas for themselves, by taking a product apart, as well as being informed of the key issues. While not necessarily directed at developing innovative ideas (although that may be an incidental by- product) the workshops are directed more at understanding the wider context of producer responsibility legislation relative to the impact this will have on the design, production and consumption of the company's products.

The intended outcome is a 'cleaner' designed product that has reduced environmental impact, while conforming to other principles of good design, i.e. it is cheaper to produce and meets market and user requirements. The product may have lower energy consumption in use, use less material in production or be more recyclable. Moreover, the product will be assessed relative to the forthcoming producer responsibility regulations.

The workshops currently focus on two categories:

1. Any products containing electrical or electronic components - equipment which is dependent on electric currents or electromagnetic fields in order to work properly…"
2. Automotive vehicles or their components

Companies receive one day's free consultancy from a DesignTrack advisor. This consultancy comprises:

- An initial telephone discussion about a particular company, product and visit details.
- A visit to the company site to analyze a product and identify re-design improvements. The visit starts with a tour of the manufacturing facility. This gives invaluable insights into how the product is made as well as the general sophistication of the company.Next a short presentation is given to members of the organization. This covers current and future legislation and is targeted directly at each company. A PowerPoint presentation gives an outline of crucial design and management issues which are then revealed and explored in the connected disassembly exercise. The presentation is supported by recent Case Studies. To get the best from the visit, a range of staff participate workshop e.g. a design engineer, marketing staff and top management.

- After some discussion, a product is disassembled.
- Key points and issues are recorded and discussed. The session finishes with a summary of the possible re-design improvements embodied in an action plan for implementation
- A follow up phone call one month or more after the visit to discuss the company's initial progress in implementing the action plan

6.3 Designtrack report

Each workshop is followed up by a short report. In each case the report contains the following information.

- The background and aim of the Designtrack visit
- The particular issues facing the company (legislation, customer needs, market position, environmental awareness)
- Analysis of the product – e.g. high-level assessment of main environmental impacts over product lifecycle)
- Design improvement opportunities
- An estimate of potential environmental and cost savings
- An action plan that the company can implement within 1 year's investment/management cycle. Any information or other requirements raised by the visit - possibly ideas for new products or even business models.

Providing a report is a good way of reminding the company of the key issues and provides a focus for cleaner design improvements with clear responsibility attributed to specific staff. The Champion has an important role as the person who will make any necessary changes. It is therefore important to assign a Champion at an early stage.

6.4 Designtrack results

Table A gives an overview of the current status of each company after a one-day DesignTrack workshop.

6.4.1 Categories

Sector: Size – Turnover (£): **Size** – No. of employees: **QS** – Quality Management Standards: EMS – **Environmental management Standards** – ISO 14001**: Take Back** – Ability to close the loop and take back their own products: **Recovery** – 'means recycling materials, component re-use or burnt with energy recovery (similar to recycle): **Recycle** - means reprocess, re-use, remanufacture any materials or assemblies: **Legislation** – is the company compliant, or even aware of the forthcoming extended producer responsibility legislation?

The companies provide an overview of issues across UK manufacturing companies relative to the financial and environmental benefits a DesignTrack workshop can bring. The companies are listed in the order of the visits, starting with the pilot companies (A-E). The focus of this paper is upon the visits undertaken by the author.

Table A: Company Status

Co.	**Sector**	**Size £**	**Size (employees)**	**QS**	**EMS**	**Take Back**	**Recovery**	**Recycling**	**Leg**
A	Medical equipment	5m	90	✓	-	-	-	-	2008
B	Medical equipment	12m	120	✓	-	-	-	-	2008
C	Electronic test equipment	0.9m	10	✓	-	-	-	-	-
D	Printed Circuit Board Manufacturer	8.9m	150	✓	✓	-	-	-	✓
E	Mobile Phones	4bn	-	✓	✓	✓	-	-	✓
F	Large Domestic Appliances	90m	800	✓	-	-	-	-	-
G	Hi-Fi equipment (consumer)	50m	20 (R&D)	✓	-	-	-	-	-
H	Railway signaling	10m	220	✓	-	-	-	-	-
I	IT & telecommunications	40m	150	✓	✓	✓	✓	✓	✓
J	Digital transmission/amplification devices	13m	36 (1200)	✓	✓	✓	✓	✓	✓
K	Printing equipment -	45m	300	✓	-	-	-	-	-
L	Medical Equipment	13m	130	✓	✓				✓
M	Lighting	0.6m	6	-	-	-	-	-	-
N	Electronic test equipment	20m	200	✓	✓	-	-	-	✓

Table B gives an overview of the benefits of a DesignTrack visit to each of the companies. While largely informal this information is constructed from post visit feedback forms and telephone interviews with participants.

Table B: Benefits of Designtrack Workshop

Co	Annual cost saving £	Time (lab)	Sub-stitution	Inno-vation	Platform	Market growth	Catalyst	Culture/ behaviour
A	£1.13m	✓	✓	✓	✓	✓	✓	✓
B	£24,000		✓				✓	✓
C	£14,000	✓	✓	✓	✓	✓	✓	✓
D	£5400						✓	✓
E	-							✓
F	12m	✓	✓	✓	✓	✓	✓	✓
G	-		✓	✓			✓	
H	£275,000	✓	✓	✓	✓	✓	✓	✓
I	-						✓	✓
J	£87,500	✓					✓	✓
K	£450,000	✓	✓				✓	✓
L	-	✓			✓		✓	✓
M	-						✓	✓
N	£82,360	✓	✓	✓			✓	✓
TOTAL £ 14,068,260 (potential savings)								

✓ A tick indicates a positive impact.

Annual cost saving (potential) – these are estimated on the basis of reduced purchase of raw materials, reduced consumption of energy, water and materials, reduced waste disposal costs, reduced labour costs and other savings e.g. avoided purchase of capital equipment, reduced distribution costs, increased re-use, remanufacturing or recycling opportunities. Total annual cost savings are estimated by calculating the cost savings per unit of product manufactured and then scaling this up based on last years production figures.
Substitution – Alternative materials with reduced environmental impact. Substitution could also mean replacing a product offering with a service – so called 'dematerialization'.
Innovation – Ideas that may lead to product/service innovation, or a new product form that can energize the product portfolio. This may be a new function, customer demand or enabling technology.
Platform–A platform product maximises the commonality of the components across the whole product range. It is a product that is part of a family of products built from a single platform of common product structures, technologies and automated product process'. (19]) The product under scrutiny could be made obsolete by considering the relationship of its component and assemblies across the whole range. This approach is exemplified in examples **A**, **C** and **F**.
Market growth – By considering the cleaner design of the product in question a new market opportunity may result. It could be the potential to energize the exiting market with new product/re-launch, added features /segmentation, or an opportunity to supply new markets.
Catalyst – the democratizing process of the workshop means ideas can be brought into the open and formalized. In some cases the ideas had previously been discussed in the company.
Culture/Behavior – 'the way we do things round here'. In a number of cases the exposure to DfE issues can influence the way companies consider product design early on in the process. This can lead to embedding DfE principles in mainstream development activity. These become transferable across company practices and team approaches to design.

SUMMARY

This paper has demonstrated the economic and environmental benefits of DfE deployed through an action workshop process.

The issues outlined in the DesignTrack workshop can instigate a fundamental shift in the way companies think about the design of their products.

This is an additional set of criteria added to technical performance, manufacturing, safety, cost, appearance, and style in product design. This has to be seen as an opportunity for improvement rather than a tiresome burden.

Product developers need to think about the 'whole of life' impact at the beginning of the process. This is not a new imperative, but now it has some legislative bite. Those enlightened firms which have taken up the DesignTrack workshops are making a significant impact on company practices.

Evidence has shown that many companies do have weak innovation management and poor creative responses to the demand for sustainability and DfE. This is particularly true at the early stages – the so-called' fuzzy front end' (20) of product development.

In their own defense, and in anticipation of even more severe legislation, businesses should ahead of the game, not just one step in front of the legislation. This means many interrelated things: a better sense of appropriate models of innovation, more finely tuned antennae to future trends, more scenario planning to guard against disruptions. Central to this is their own grasp of specific technological and material impacts .By including environmental factors early in strategic planning, companies are better able to pre-empt regulatory changes and thus better manage the potential risks of industrial operations and products downstream.
In order to proactively defend themselves, businesses should be thinking harder and doing more.

REFERENCES

(1) Less without loss' was an expression first used in the co-design - The Interdisciplinary Journal of Design and Contextual Studies. Editorial of 03/1996 by David Walker, Phillip Goggin and Eric Billet.
(2) Shabi, R., "The E-Waste Landscape", The Guardian Weekend, 2002 Nov 30th
(3) Environment Agency – Bristol
(4) Goggin, P, A,. Glossary: Key Terms and Definitions. Co-Design. The Interdisciplinary Journal of Design and Contextual Studies. (1996)
(4) Fiskel, J. (Ed) Design for Environment, McGraw-Hill, New York, (1996)
(5) Jonathan Rosenhead and Jonathan Mingers, Rational Analysis For A Problematic World Revisited, Wiley (2001)
(6) Thackara, J. Winners How Today's Successful Companies Compete by Design. Gower (1997)
(7) Senge, P,. The Dance of Change. The Challenges of Sustaining Momentum in Learning Organisations. Nicholas Brealey (1999)
(8) William McDonough and Michael Braungart, Cradle to Cradle, North Point Press. (2002)
(9) Gracdel, T.E., and B.R. Allenby Industrial Ecology Prentice Hall (1995)

(10) Thackara, J. Winners How Today's Successful Companies Compete by Design. Gower (1997)
(11) See www.dti.gov.uk/sustainability
(12) Fridge Recycling: A Technical Insight, EMR Limited, www.emrltd.com
(13) Royal College of Art, Dept. Industrial Design Engineering www.ide.rca.ac.u.k
(14) Cambridge University, Institute for Manufacture and Management and Engineering Design Centre.
(15) Holdway, R., Fraser, P., Moultrie, J., Designing Better Business: Assessing & Strengthening UK SMEs Design Capability: principles into Practice, DMI Boston. The 11th International Forum on Design Management Research & Education Northeastern University (2002)
(16) Pfeffer, J,. and Sutton, I. R., The Knowing Doing Gap. How Smart Companies Turn Knowledge into Action. Harvard Business School (2000)
(17) Simonsson, A., and Barthel, M,. Organisational aspects of the application of Sustainable Product Development and Design, Sustainability Group, British Standards Institution (BSI), UK(2002)
(18) Anecdotal evidence described by Design Council in conversation with author. (March 2003).
(19) Meyer, H. Marc and Lehnerd, P. Alvin The Power of Product Platforms. Free Press. (1997)
(20) Smith, G., P., and Reinertsen, G., D., Developing Products in Half the Time. Van Nostrand Reinhold (1991)

Acknowledgement
The Good Design Practice Project is supported by the Monument Trust

Product design education in practise – evaluating the key transition from undergraduate degree to initial industrial position

D EGGBEER, J REES, P DORRINGTON, H MILLWARD, and **A LEWIS**
The National Centre for Product Design and Development Research (PDR), University of Wales Institute, Cardiff, UK

ABSTRACT

This paper assesses the educational and training needs of recent product design graduates by evaluating their progress from undergraduate degree through to their first two years in industry. A case study approach is used to track three graduates from UWIC's Product Design degree course through to employment within TCS (formerly known as the Teaching Company Scheme) programmes operating in small and medium-sized enterprises (SMEs). The undergraduate education and the structured TCS-based training are reviewed, and the challenges associated with working with SMEs are highlighted. Generic educational and training issues are identified, with particular emphasis on design-for-manufacture and ethical design.

1. INTRODUCTION

Product design is a key component in the new product development cycle, and hence an essential component within the wealth-creation process and the economy in general. In order to maximise the effectiveness of the product designer within industry, consideration needs to be given to two key areas: (a) education received at university degree level; and (b) vocational training received in the industrial sector. Novice product designers entering the workplace rely on a solid undergraduate education to prepare them for the challenges of industry, and a number of approaches have been postulated for promoting industry-relevant skills (1). Within the work-place environment, there is a wide variety of literature focused on developing and refining product design techniques, such as systematic design and stage-gate processes (2). However, the majority of product design education studies report industrial collaborations with large well-established companies (3). There has been limited attention paid to analysing the needs of young product designers as they undertake the key transition from undergraduate degree to initial industrial position within small companies, specifically small manufacturing companies who often lack a clear understanding of design.

SMEs represent an important element in national economies around the world, and they play a significant role in the design, development and manufacture of new products. In the EU an SME is categorised as a company employing fewer than 250 people. In fact, more than 95% of the three million businesses in the UK employ fewer than 20 people (4). The majority of the PDR-based TCS programmes have focused on introducing design technology and techniques into 'traditional' manufacturing SMEs. Advanced design technologies, such as computer aided design (CAD) software, can bring significant benefits to SMEs (5). However, the idiosyncratic nature of small companies means that this can be a culture shock for young product designers. A number of research studies have highlighted the difficulties of implementing new product development activities within SMEs (6). Furthermore, senior management within SMEs seldom have the education or training relevant in product design activities. Within the manufacturing industry, there is a tendency for management to attempt to cultivate design literacy 'through osmosis' rather than through formal targeted training (7), and this has an impact on training standards for subordinate staff.

This paper will assess the educational and training needs of recent product design graduates by evaluating how they coped with the transition from undergraduate degree through to their two-year TCS programme within a manufacturing SME. The SME environment is not necessarily conducive to ethical design but the TCS programmes are structured and closely monitored, and aim to develop both the technical and personal competencies of the graduates. Therefore this research, based on a series of case studies, will provide a unique insight into the particular challenges encountered by product designers working within SMEs. The aim is to highlight key degree course elements that support their transition into industry, and also to identify work-based training requirements. In addition, the case study material will be used to show the extent to which the TCS programmes prepare the graduates for education at the research degree level.

2. TCS MODEL

PDR have employed the TCS model as an effective mechanism for partnership and collaboration with a wide range of SMEs. TCS has been in operation for over 20 years, and is a government-backed knowledge transfer scheme. The aim of the scheme is to strengthen the competitiveness and wealth creation of the UK by stimulating innovation in industry through structured collaborations with universities and research organisations. The programmes are usually two years duration and provide employment for a well-qualified graduate TCS Associate for the duration of the programme.

All the PDR-based TCS programmes are focused on product design, and the numbers reflect the UK trend in that the majority have been based in small companies. PDR have successfully completed nine TCS programmes since 1995, and there are ten PDR-based TCS programmes currently ongoing. In line with other researchers (8), PDR have found that the TCS model is an ideal vehicle through which to evaluate the training needs of product design graduates, and assess the application of these skills within new product development activities.

The well-defined management and structure of the TCS process promotes a detailed analysis of both the TCS Associate and the company from the university partner's perspective. PDR is based within UWIC, and have found that the BSc Product Design course at UWIC provides a good foundation for a TCS Associate. Once an appointment has been made, the Associate is assigned at least two PDR-based supervisors. Regular contact with the TCS Associate and company fosters a level of trust and co-operation that generates an in-depth understanding of the subtle issues and problems inherent in any small company. TCS programmes are characterised by a commitment to disciplined effective project management through mandatory monthly and quarterly meetings. The monthly meetings between the supervisors and the Associate focus on the technical issues within the programme, as well as addressing training and personal development requirements. As part of the TCS programme, each Associate attends four separate one-week training modules, and it is estimated that 10% of their time is spend on training-related activities.

This research study has selected three anonymous TCS Associate case studies to illustrate a range of product design education and training issues, specifically within the distinctive working environment of a manufacturing SME. The selection criteria for the case studies were as follows:

a) Good honours graduate from UWIC's BSc Product Design course;
b) Completed or undertaking a PDR-based TCS programme with a small company, of between 10 and 50 employees;
c) Employed within the 'traditional' manufacturing sector, with a commitment from within the company to enhance its design capability.

3. UWIC's PRODUCT DESIGN COURSE

The undergraduate course is a three-year full-time course, providing a BSc Honours degree accreditation. The course runs live projects with industry, but does not include formal work placements. However, each of the three TCS Associates selected for this research study completed periods of relevant work experience during the degree course. The course aims to combine the traditional attributes of product design and mechanical engineering, with particular emphasis on the design engineering skills required during the development of a new product. A summary of the three-year course is given below.

Year 1 Modules (100% taught)	Year 2 Modules (100% taught)
The Marketing & Design Interface	Marketing Influences on Design
Design Models & Methods	Information Ergonomics
Ergonomics in Design	Concept Generation & Development
Effective Communication of Design Concepts	Design for Manufacture & Assembly
Workshop Practice & Model Making	Manufacturing Analysis & Reverse Engineering
Materials & Manufacturing Process Selection	Integrated Design & Concurrent Engineering
Technical Specifications	System Control & Instrumentation
Engineering Science for Product Designers	Mathematics 2
Electronics for Product Designers	Mechanical Engineering Studies A
Mathematics 1	Mechanical Engineering Studies B
History of Design & Technology	CAD Software Training: I-DEAS
Computer & IT Studies	Computer Aided Mechanical Engineering
Year 3 Modules (15% taught, 85% project-based)	
Business Management & Professional Practice	
Project Management & Product Development	
Advanced Design Option: self-directed, research-based product/technology project	
Major Project: self-directed design project	

The Year 3 major project was an important element within the course, whereby the students developed a design principle from initial concept through to final working prototype. This element brought together their 3D CAD, rapid prototyping and model-making skills. It should be noted that the opportunities to develop these skills varied between students. For example, one Associate was only introduced to basic CAD solid-modelling, whereas another chose to make extra use of 3D CAD and was effectively self-taught.

4. TCS ASSOCIATE CASE STUDIES

4.1 TCS Associate A: Safety equipment manufacturer

Associate A was employed in a small family-run company that specialised in the manufacture of safety and survival equipment for the outdoor and military markets. The company were active in identifying new market opportunities, but required a structured development process to bring their new products to market. The TCS programme was configured to address this need. However, the company was sales-orientated; it lacked adequate market research and initially resisted the concept of structured innovation. Therefore, the challenge for Associate A was to introduce new working practices that could be understood and applied by a range of different staff.

Product designs were brought to realisation through the use of a 3D CAD workstation employing I-DEAS software. New product designs were successfully developed using a structured stage-gate process, with formal sign-off and design reviews to approve and monitor new projects. In parallel, a product development training programme was implemented across the company. A summary of the training undertaken by Associate A during the TCS programme is given below.

TCS Modules	(1) Developing Project Handling Skills; (2) Improving Personal Skills; (3) Commercial and Technological Change; (4) Changing Business Environment.
TCS Mini Project	'Designing and developing a survival item'
Professional Training	Process FMEA; Marketing for Beginners; Ergonomics; Aesthetics; Design Analysis; NVQ Management; Sustainability in Design; Anatomy and Physiology; Rehabilitation Engineering; Photography.
Informal Training	Meetings with tool makers, project stakeholders and industry experts (specifically manufacturing technology and techniques); Finite Element Analysis (FEA); MoD Standards; Macromedia Director; Adobe Photoshop.
MPhil Degree	Withdrew from higher degree study programme

4.2 TCS Associate B: Injection moulding company

Associate B was employed in a small 'traditional' injection moulding company. Their main customer base was the automotive, electrical and electronics industries. However, in order to generate strategic growth, the company needed their own in-house design capability and to explore new high-added-value markets, specifically medical mouldings. The TCS programme facilitated this major change in company structure and direction.

Associate B implemented an I-DEAS 3D CAD workstation to drive the in-house design and development activities. In parallel, a significant amount of work was put in to ensuring that the company systems conformed to the rigorous medical standards required for product development and clinical trials. Working within accredited standards demonstrated that the company was fully committed to developing high-quality medical products. In addition, Associate B implemented and upgraded the company's quality system (ISO 9001) and environmental management system (ISO 14001). A summary of training undertaken by Associate B is given below.

TCS Modules TCS Mini Project	[identical to Associate A] 'Material utilisation and waste minimisation'
Professional Training	CAD/CAM modules from UWIC's BEng Mechanical Engineering course; Institute of Environmental Managers Association (IEMA) (1) Foundation, (2) Associate Membership and (3) Internal Auditor courses; NEBOSH General Certificate in Occupational Health & Safety.
Informal Training	Presentation techniques; Value Stream Mapping; Benchmarking; Statistical Process Control; Design FMEA; FDA & MDA medical device regulatory requirements; Sterilisation techniques; Rapid prototyping techniques; ISO 14001; Investor in People.
MPhil Degree	Title: 'Implementing Product Design within an SME'. Completed UWIC's Certificate in Research Methods; MPhil ongoing.

4.3 TCS Associate C: Mass transit supplier

Associate C was employed by a small manufacturing company that supplied a wide range of components and assemblies for the mass transit sector. Fabrication was based on one of three core in-house technologies: (a) metal alloy castings; (b) polyurethane mouldings; and (c) composite mouldings. The TCS programme was established to implement a knowledge-based in-house design capability, such that essential manufacturing information and constraints could be integrated into a customer's design much sooner in the development cycle. The company aimed to increase their new product development activities and gradually make the transition from 'just manufacture' to 'design and manufacture' contracts.

A 3D CAD workstation, running I-DEAS software, was installed early in the TCS programme. This acted as a very effective customer interface through which all design communications and decisions could be channelled. Associate C then linked this design platform to a 5-axis CNC machining centre, using the appropriate I-DEAS CAM software. This design-to-manufacturing link allowed the company to capture major rail contracts and established a reputation for high-quality, consistent composite products. A summary of training undertaken by Associate C is given below.

TCS Modules TCS Mini Project	(1) Developing Project Management Skills; (2) Improving Personal Skills and Teamwork; (3) Business Management Skills; (4) Exploring Business Opportunities. 'Reducing waste in the aluminium casting process'
Professional Training	Design FMEA; Design with Composite Materials; NVQ Management; Adobe Photoshop; First Aid at Work; Italian.
Informal Training	CAM training with PDR personnel; 3D CAD Surface Modelling tutorials; Solid Modelling seminars; questionnaire design and interview techniques.
MPhil Degree	Title: 'An evaluation of the commercial and operational impact of implementing design-led technologies within a manufacturing SME'. Completed UWIC's Certificate in Research Methods, conference paper accepted; MPhil ongoing.

5. DISCUSSION

The UWIC Product Design degree represents the common educational element in this paper. The case studies indicate that the degree course provided each of the three Associates with the fundamental technical skills necessary for new product design. However, this level of education alone is insufficient to manage the full design and development cycle within the context of a graduate's first industrial placement. The degree course provided the graduates with an appreciation of a wide range of design tools and techniques, and the key benefits of the three-year degree can be summarised as follows:

- Introduction to 3D CAD;
- Knowledge of model-making and rapid prototyping techniques;
- Good understanding of design techniques, ergonomics and market research principles;
- Introduction to technical/material specifications and manufacturing technologies.

The implementation of 3D CAD within a 'traditional' manufacturing company was an important early deliverable within the three case studies. It is significant that the I-DEAS software taught at university level was the chosen software in each of the three TCS programmes. The benefits of 3D CAD software are numerous, but it would appear that personal choice and experience heavily influence the CAD selection process. Although the CAD training at university level was sufficient to initiate this software-based design technology, two of the three Associates required further CAD training as part of their TCS programmes. This was particularly evident for Associate C, who had the responsibility of combining CAD and CAM software to drive the company's high-quality machining operations. The university degree did not cover this crucial link between design and manufacturing software. In addition, each Associate undertook Failure Mode and Effects Analysis (FMEA) training. This risk assessment technique was clearly seen as an important element in any new product development process.

The case studies showed a shortcoming within the university degree was a lack of focus on the important design-to-manufacture interface. Within the context of small manufacturing companies, each of the three Associates considered that their degree did not adequately prepare them for design-for-manufacture activities. This is a broad area covering CAD/CAM/CAE integration, appropriate materials selection, tooling design and CNC machining. It would appear that there are no 'ready made' training modules whereby a young product designer can quickly acquire these key skills and, as such, they have traditionally been addressed through industrial placements and learning through work experience. With this in mind, the following additions should be considered for the UWIC degree course:

- Industrial placement(s), with emphasis on design-for-manufacture;
- Company case studies to highlight the industrial perspective on design;
- External lecturers from industry – a review of industrial practices (both good and bad) and what industry is looking for in a graduate;
- Management skills – improving communications, facilitating company change and implementing internal audits;
- Introduction to ethical design and environmental/sustainable design issues.

The three Associates reported that ethical design represented an important gap within UWIC's product design syllabus. Across the three TCS programmes, ethical design manifested itself predominantly through environmental considerations. In order for the Associates to contribute to this important subject, extra training was necessary. Associate B undertook three separate environmental management courses, which directly helped to implement world-class ISO14001 standards, and this had a significant impact on the TCS programme. In addition, Associate A attended a dedicated sustainable design training course, and Associate C implemented cost reductions through a waste reduction project. However, the key barrier to implementing effective ethical design was not the lack of education and training, but rather it was the idiosyncratic nature of SMEs themselves.

The main challenges faced by the three TCS Associates in implementing new product development were due to the difficulties associated with working in SMEs (6). Manufacturing SMEs tend to be resource-constrained environments in which senior managers focus on reducing time and costs at the expense of product quality and consistency. Education and training can address technical and practical skills, but in this type of working environment communication and interpersonal skills are of equal importance in order to overcome the main barriers to new product development. These barriers can be summarised as follows:

- Dominant owner/managers with a lack of understanding of design, e.g. continual intervention without their full appreciation of the project;
- Inadequate market research and unrealistic expectations with regard to new product development time and costs;
- Failure to understand the need for a structured development process and a resistance to adopt appropriate design documentation;
- Short-term, 'fire fighting' strategies encouraged rather than long-term strategic planning;
- A sales-driven ethos at the expense of a robust strategy which integrates business with design.

The fact that successful results were delivered in each of the three TCS case studies is due largely to the commitment and perseverance of the Associates, and the structure and discipline imposed through the university collaboration. The university support team, in combination with the four TCS training modules, ensured that the Associates were given the time and opportunity to develop their technical, managerial and personal attributes to a much higher level. This support afforded an acceleration of the learning process beyond undergraduate degree level that would have been far more difficult had the graduates been alone and isolated in a typical manufacturing SME. The combination of professional and informal training elements undertaken during the TCS programmes enhanced the technical knowledge developed at university level and expanded this within a business context. Consequently, the Associates received a greater understanding of product design with a small company, and learned how to implement effective design and development within the distinctive environment encountered within a manufacturing SME. The TCS-based training also broadened the graduates' education in terms of management skills, best practice business strategies, communication skills and customer-relation techniques.

The university is an integral part of the TCS programmes, and the academic teams encourage and support education at the higher degree level. Many of the projects within the TCS programmes provided opportunities to undertake research that would be directly applicable to Masters level degree. However, none of the three Associates completed their MPhil degree within the two-year programme period. In fact, only one Associate actively generated research data throughout the duration of the programme, and then continued to pursue the MPhil during his own free time. It would appear that the workload and challenges associated with SME-based TCS programmes are too great to be able to accommodate a MPhil degree. The priority for the Associates was to successfully implement new product development, therefore any Masters level degree should be able to accredit this significant amount of work towards the higher degree. Other universities have developed an MSc degree by Learning Contract (8), and this work-based learning model should be considered for future TCS developments.

6. CONCLUSIONS

UWIC's Product Design degree course provides a solid foundation for the application of new product development; however, no three-year degree syllabus can fully prepare young graduates for coping with the unique challenges of working in small manufacturing SMEs. The main deficiency at university level was a lack of training and insight into design-for-manufacture. Greater industrial perspective and targeted educational modules can address this. Within the industrial sector, there is a need to promote systematic design-for-manufacture training, which can compliment the focus on CAD/CAM software technology. The TCS model provides an effective mechanism for continued professional development post-university, and the support structure ensures that training is maintained and expanded, even though the graduates may well be working in resource-constrained restrictive environments. These environments are not conducive to ethical design, therefore this area needs to be addressed within educational and training modules, specifically encouraging ecological and sustainable design issues. Integrating design-for-manufacture and ethical design within a manufacturing SME is a significant challenge, therefore it is PDR's recommendation that a TCS programme should be the chosen route for small companies wishing to employ a young product design graduate. Future work in this area should examine work-based learning models that successfully incorporate higher degrees into two-year TCS programmes.

REFERENCES

(1) Martin A., Milne-Holme J., Barrett J., Spalding E. and Jones G. *'Graduate satisfaction with university and perceived employment preparation'*. Journal of Education and Work 13(2) 2000, 43-41.

(2) Cooper R. *'Stage-gate systems: a new tool for managing new products'*. Business Horizons (May-June), 1990, 44-54.

(3) MacDonald A. and O'Neil L. *'In partnership with industry: the educational-industrial interface'*. Journal of Art and Design Education 16(1) 1997, 25-34.

(4) Cawood G. *'Design and innovation culture in SMEs'*. Design Management Journal 8(4) 1997, 66-70.

(5) Brown R., Lewis A. and Mumby A. *Enhancing design effectiveness in very small companies*. Design Engineering Research Centre, Cardiff, UK. ISBN 1901357007 (1996).

(6) Lewis A. and Filson A. *'Cultural issues in implementing changes to new product development processes in an SME'*. Journal of Engineering Design 11(2) 2000, 149-157.

(7) Formosa K. and Kroeter S. *'Toward design literacy in American management: a strategy for MBA programs'*. Design Management Journal 13(3) 2002, 46-52.

(8) Lipscomb M. and McEwan A. *'The TCS model: an effective method of technology transfer at Kingston University'*. Industry & Higher Education (Dec) 2000, 393-401.

Inclusive product design (ethics and sustainability) project teaching, using a major study project as the vehicle

C R GENTLE, A R CRISP, and **J LORD**
School of Engineering, The Nottingham Trent University, UK

1. INTRODUCTION

The author has previously described (Crisp et al 1999) the theoretical implementation of ethical, sustainable and environmental issues within the engineering curricula by the School of Engineering at the Nottingham Trent University. This paper describes both the continuing design and development of a tricycle 'running frame' together with a parallel attempt at pragmatic ***'inclusive design'*** implementation within the syllabus via the major study project. The author (Crisp) has over a period of six years researched the area of support mechanisms for disabled athletes, which has led to the design, development and production of a running frame for both athletic and social use by the disabled. Initially aimed at the European market, it is now seen as a viable alternative mode of transport for the disabled throughout the third world. Recent developments have focussed on the re-use of ***'end of life'*** cycle components for the frame, which has resulted in two further pathways of development, one, reuse for the building of ***'tricycles'*** and two, reuse for the building of ***'suspension frames'.***

2. THE PROGRAMME OF STUDY

The BA/BSc Honours Product Design Programme was validated and recruited to in July of 1999 and September 1999 respectively. The original programme was taught in collaboration between the School of Art and Design and the School of Engineering, within the Faculty of Engineering and Computing. It was perceived as a truly integrated programme built on the bedrock of expertise drawn from the broad spectrum of activity which is design i.e. staff from art, design and technology. The programme was seen as befitting the aspirations of the 'Great Exhibition' and Prince Albert's coupling of the twin peaks of creativity that are art and science. It is interesting to note that the Wellcome Foundation is currently asking the question' Science and Art; Symbiosis or just good friends?' (1)

2.1 Original Programme of Study

The original programme of study saw commonality across the pathways until the second half year of the second year of study, when the decision to follow BA or BSc pathway had to be made. The programme was modular and taught across a full academic year. The modules in the first year are seen as introductory and necessary for the student to make an informed judgment on the pathway of study.

Technology was taught in discrete modules separate to design project work but with an integrating theme, which ran through the assessment of both modules. Design studies was taught to all students for the first and second years, Graphical methods and CAD taught in the first year and management and marketing taught throughout the second year. The final year was differentiated by two modules of technology for the BSc pathway and a minor project and dissertation for the BA pathway.

2.2 Existing Programme

Table 1. Existing Programmes of Study for Product Design

BSc Honours Product Design	BSc Product Design H715	BSc Computer Aided Product Design H131	BSc Product Design: Sport and leisure W243
Year 1, First Half, 60 points common	Introductory Design Technology Project, 40 Credit Points	Introductory Design Technology Project, 40 Credit Points	Introductory Design Technology Project, 40 Credit Points
	Introduction to Design Studies, 10 Credit Points	Introduction to Design Studies, 10 Credit Points	Introduction to Design Studies, 10 Credit Points
	Graphical Methods 1A, 10 Credit Points	Graphical Methods 1A, 10 Credit Points	Graphical Methods 1A, 10 Credit Points
Second Half, 40 points common	Group Design Technology Project, 40 Credit Points	Group Design Technology Project, 40 Credit Points	Group Design Technology Project, 40 Credit Points
	Design Studies 1, 10 Credit Points	Introduction to 3 D CAD, 10 Credit Points	Human Factors in Design, 10 Credit Points
	Graphical Methods 1B, 10 Credit Points	Computer Aided Conceptual Design, 10 Credit points	Introduction to Biomechanics, 10 Credit Points
Year 2, First Half, 50 points common	Industrial Design Project, 40 Credit Points	Industrial Design Project, 40 Credit Points	Industrial Design Project, 40 Credit Points
	Management and Marketing, 10 Credit Points	Management and Marketing, 10 Credit Points	Management and Marketing, 10 Credit Points
	Design Studies 2A, 10 Credit Points	CA Surface Modelling in Design, 10 Credit Points	Human Physiology, 10 Credit Points
Second Half, 50 points common	Computer Aided Design Project, 40 Credit Points	Computer Aided Design Project, 40 Credit Points	Computer Aided Design Project, 40 Credit Points
	Finance and Quality Management, 10 Credit Points	Finance and Quality Management, 10 Credit Points	Finance and Quality Management, 10 Credit Points
	Design Studies 2B, 10 Credit Points	Multi-Media Portfolio development, 10 credit Points	Biomechanics, 10 Credit Points
Year 3, two pathways common, share 90 points with third, 30 with the fourth.	Major Study Project with Integrating Thesis, 60 Credit	Major Study Project with Integrating Thesis, 60 Credit	Major Study Project with Integrating Thesis, 60 Credit
	Design Technology, 30 Credit Points	Design Technology, 30 Credit Points	Design Technology, 30 Credit Points
	Computer Aided Technology, 30 Credit Points	Computer Aided Technology, 30 Credit Points	Sport Science Technology, 30 Credit Points

The original programme was developed throughout the academic years of 2000-2001 and 2001-2002 as follows,

- Technology integrated with the project modules,
- The year divided into two

- Each half year containing one forty and two ten credit point modules
- Appropriate CAD taught in all project modules i.e. Pro-Engineer/3 D Studio Max
- Pathways added to the programme for a Computer Aided product Design and
- Product Design: Sport and Leisure
- The pathways defined as separate programmes of study for the September 2003 recruitment.

Within certain modules specific attention was paid to timely concerns of both professionals and the society in general. These concerns are often focussed under the generic terms of environmental issues or inclusive design. These issues are then discussed to take on board such theories and terms as sustainability and subsidiarity. The main themes incorporated into the teaching curricula are those exemplified by the schools council of 1974 and drawn together by Huckle 1991 and Sterling/EDET 1992 (2). These issues are discussed and promoted to the forefront of the students thinking by specific modules e.g., Design Studies, where the process is explicit and through the design project modules, where the process is implicit.

Table 2. Design Studies Curricula Content

Design Studies 1 (20 CP)	Programme Specific Learning Outcomes
• Historic and biographic studies • Design as a context-based activity • Design, its role in innovation • Political, social, economic and technological impacts on design	• Demonstrate an awareness of contemporary design issues and themes • Demonstrate an awareness of design in its surrounding context
Design Studies 2 (20 CP)	Programme Specific Learning Outcomes
• Design responsibilities and integrity • Life cycle analysis • Environmental health and safety legislation • Consumerism and its effects • Patents	• Understand moral and ethical responsibilities of design (e.g. the concept of sustainability). • Knowledge of legislation affecting product design. • Knowledge of the rights of the public/consumer. • Knowledge of intellectual property rights and protection in the market place.

Table 3. Major Study Project Curricula Content

Major Projects (60 CP)	
• Taught element, the design process • Project, including an underpinning thesis	• Ability to research for product designs and apply an appropriate model of the design process in development and testing. • Demonstrate the ability to generate and interpret computer simulations for evaluation and testing • To take account of costing • To apply appropriate skills with academic rigour • Ability to produce interactive computer presentations • Integration of principles taught in Design Studies • Authoring of supportive thesis • Realisation of the Artefact

3. THE MAJOR STUDY PROJECT

The Product Design Specifications (PDS) for the students major study project incorporated many of the issues relevant to the philosophy of design as taught and expressed currently by academia. In particular when assessing this project and in detail the technical aspects of the three designs and integrated development, comment is also made on the influence on the design by contemporary issues concerning ethics, morals, life cycle analysis, inclusive design and sustainability e.g., did the design seriously integrate these issues and indicate a thorough understanding of the issues by the student.

3.1 The product design specification

The students PDS contained many objectives, with those pertinent to this paper as follows;

- Design and manufacture a running frame from recycled parts
- Mainly for use by sufferers of Cerebral Palsy
- Cantilever arms to be removable not fixed
- Dual design for suspension bikes/trikes from recycled parts
- Demonstrate an understanding of life cycle analysis and sustainability
- Take an eco-centric or techno-centric approach
- Find a compromise between economic and sustainable influences
- Approach the learning and teaching correctly
- Socially make the product totally inclusive

3.2 Sustainability

Sustainability is now recognised as a policy permeating both Global and European economic planning. However, as the European Union favours the implementation of its sustainable policies via methods of subsidiarity, normally manifesting itself within the European Union in the collection, separation and recycling of domestic, industrial and hazardous waste, this theme is explored in connection with the design of the frames.

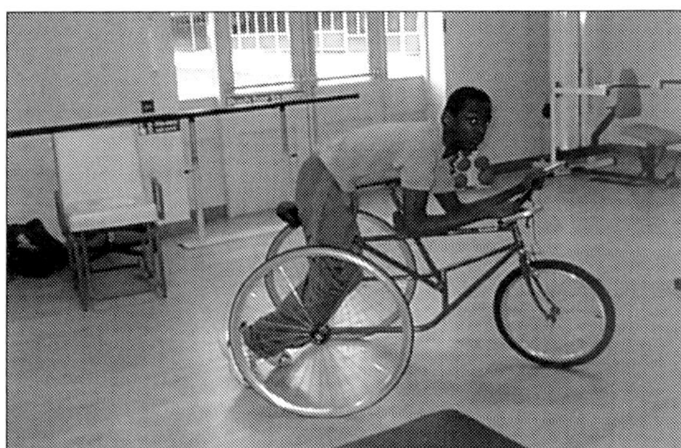

Figure 1. Cerebral palsy user

This promotion of 'thinking globally, acting locally' is ideally suited to third world countries where often both the hand skills and materials are found together and where an abundance of discarded transport components are readily available for re-use.

4. THE EXISTING RUNNING FRAME

The running frame was originally designed for use by suffers of cerebral palsy, Figure 1; however, it soon became apparent that all ages and all conditions found the frame a pleasurable mode of transport. The running frame resembles a large tricycle, which in its original format was devoid of a driving mechanism i.e. chain, pedals and sprockets to bottom bracket and rear wheel. It is propelled along when used by cerebral palsy suffers by a 'scooting' action of one or both feet. The rear wheels, two off, are cantilevered from the frame on two rear arms allowing access for a wheel chair. Steerage is by conventional front forks and handlebars, the user resting their body mass on a large chest support, braking was originally conventional i.e. standard front calliper. It can be powered for leisure users by battery or conventional chain drive.

Figure 2. Bottom bracket utilized as suspension unit.

The desire for a more comfortable ride initiated a new line of development and indeed thought, leading the authors to use an existing bottom bracket as the fulcrum point for a suspension unit for the cantilevered rear wheel arms, Figure 2. This was followed by the design and development of a tricycle utilizing two bottom brackets (one forward of the norm) and the subsequent development of rear suspension for a normal frame again using two bottom brackets (one rear of the norm) all making use of recycled component parts.

5. FINAL DESIGNS AND MANUFACTURE OF PROTOTYPES

The student investigated the feedback from current users and determined that the main area of concern from a user's viewpoint was the chest support. The design now incorporates a full chest support, which fans out from the frame to also support the upper arms. To accommodate the removal of the cantilever arms a bottom bracket was utilised, which was dependent on cotter pins for securing the axle and cranks. The cranks being modified to fit the internal diameter of the cantilever arm prior to welding. At the same time designs were developed to enable the bottom brackets to be manufactured in pairs, back to back, this enables one to be utilised as a suspension pivot and the other as the crank bearing. From earlier developments it was found that the production of the frame was time consuming and costly, mainly a result of low volume production. The student addressed this problem by recycling two different frames, purchased for little cost second hand. A traditional 21-inch 'ladies frame' was modified as follows;

- Removal of rear chain stays
- Removal of seat pin tube
- Removal of front forks, this included removing half of the handle bar stem tube in the vertical direction
- A traditional mountain bike frame was modified as follows;
- Cut from the main frame the front forks and handle bar stem tube

These two parts were now welded together to form a recycled running frame. The cotter pin style cranks were ground and turned to make concentric bosses, fitted to the cantilever arms, which were produced from new steel tube and aligned at 180 degrees to each other before final assembly and welding. The damping unit was manufactured from recycled inner tubes, contained within a newly designed housing. The student moved the radius arms closer to the bottom bracket to stiffen the support pin running through the centre of the damping unit. The final design as realized for the running frame is shown in figures 3 and 4. The double bottom bracket design has at the time of writing progressed no further than the drawing board stage, however the success of the recycled innovations associated with the running frame suggest a similar successful out come.

Figure 3. Detail of cantilever arm and bracket assembly

Figure 4. Final realised design with full body support

6. CONCLUSIONS

The student's final design and prototype matched well his expectations and the detailed PDS. He delivered a design solution from both a technical and social perspective. The running frame can be genuinely built from recycled components with no requirement for high technology input any where around the globe. He has integrated the themes of sustainability and inclusive design particularly well and has a greater awareness of others' needs from doing so. He has also rediscovered 'old technology' to be of great benefit e.g., the humble cotter pin. However, from a tutor's viewpoint, he has demonstrated that complex social issues can be addressed whilst also addressing product design requirements and the personal goal of obtaining a worthwhile degree. He will leave the university with a legacy of acceptable designs and the knowledge the programme is valid, he himself will leave a broadly educated person with the flexible technical and social skills to pursue a worthwhile career and be able to tap into at any future point another programme of lifelong learning.

REFERENCES

(1). Wellcome News, Supplement 5, Q1 2002 ISSN 1356-9112

(2). Palmer J *A "Environmental Education in the 21st Century* "Routledge 1998 London. ISBN 0-415-13197-9 pp. 139

(Crisp et al 1999) ENTRÉE PROCEEDINGS ISBN 90-76760-01-2 pp 217

Authors' Index